National Detector Dog Manual

United States Department of Agriculture

Marketing and Regulatory Programs

Animal and Plant Health Inspection Service

Plant Protection and Quarantine

Fredonia Books
Amsterdam, The Netherlands

National Detector Dog Manual

by
United States Department of Agriculture

ISBN: 1-4101-0819-8

Copyright © 2005 by Fredonia Books

Reprinted from the 2003 edition

Fredonia Books
Amsterdam, The Netherlands
http://www.fredoniabooks.com

All rights reserved, including the right to reproduce
this book, or portions thereof, in any form.

Acknowledgments

We would like to recognize the following individuals who worked diligently on this final version of the USDA National Detector Dog Manual:

Bruce Attavian, Technical Manuals Writer, PDC, Frederick, MD

Lisa Beckett, Training Specialist, NDDTC, Orlando, FL

Debra Dunn, WR Senior Regional Program Manager, Fort Collins, CO

Sue Ellis, NDDTC Instructor, Orlando, FL

Becky Frasure, Editorial Assistant, PDC, Frederick, MD

Roger Holman, ER Regional Program Manager, Raleigh, NC

Grace Nagano, WR Regional Canine Program Coordinator, Honolulu, HI

Alison Pae, ER Regional Canine Program Coordinator, Atlanta, GA

Frank Ramos, Administrative Support Assistant, NDDTC, Orlando, FL

Albert Roche, ER Regional Canine Program Coordinator, San Juan, PR

Michele Sowell, WR Regional Canine Program Coordinator, Houston, TX

Diana Verity, WR Regional Canine Program Coordinator, Los Angeles, CA

Jay Weisz, Director, NDDTC, Orlando, FL

Donna L. West, National Detector Dog Program Manager, Washington, DC

We would also like to ackowledge those individuals who worked on the interim manual:

Jim Armstrong, Darren Bartnik, Wendy Beltz, Dr. Rose Borkowski, Berkie Brown, Kim Caudle, Lisa Davis, Debra Dunn, Becky Frasure, Beverly Gordon, Brendt Heldt, Tony Maki, Tom Miller, Grace Nagano, Albert Roche, Andy Rodriguez, Sandy Seward, Calvin Shuler, Jim Smith, Mike Smith, and Dawn Wade.

Contents

Tables
page v

Figures
page vii

Introduction
page 1-1-1

Procedures
Equipment   *page 2-1-1*
Incidents   *page 2-2-1*
Kenneling Requirements   *page 2-3-1*
Procuring Detector Dogs   *page 2-4-1*
Public Awareness   *page 2-5-1*
Retiring Detector Dogs   *page 2-6-1*
Utilizing Detector Dogs   *page 2-7-1*

Health Care
Anatomy   *page 3-1-1*
Diseases and Parasites   *page 3-2-1*
First Aid and Emergency Care   *page 3-3-1*
General Care   *page 3-4-1*
Grooming   *page 3-5-1*

Training
page 4-1-1

Glossary
page 5-1-1

Appendix A
APHIS Contacts   *page A-1-1*

Appendix B
Personnel   *page B-1-1*

Appendix C
History of Detector Dog Programs   *page C-1-1*

Appendix D
Equipment   *page D-1-1*

Appendix E
Shipping and Daily Transporting Detector Dogs   *page E-1-1*

Appendix F
Weight Rating   *page F-1-1*

Appendix G
Expressing Anal Glands   *page G-1-1*

Appendix H
Reporting and Documentation   *page H-1-1*
Appendix I
Manual Maintenance   *page I-1-1*
Appendix J
Legislative Authority   *page J-1-1*
Appendix K
Forms   *page K-1-1*
Index
page Index-1-1

Tables

TABLE 2-4-1: Steps of the Procurement Process   *page 2-4-4*

TABLE 2-4-2: Action to Take After Completing the Initial Screening   *page 2-4-7*

TABLE 2-4-3: Action to Take After Completing the Temperament Evaluation   *page 2-4-8*

TABLE 2-4-4: Action to Take After a Veterinarian Completes the General Exam   *page 2-4-9*

TABLE 2-4-5: Action to Take After a Veterinarian Completes the Heart Worm Test   *page 2-4-9*

TABLE 2-4-6: Action to Take After a Veterinarian Completes the Blood Test   *page 2-4-9*

TABLE 2-4-7: Action to Take After the NDDTC Veterinarian Views the X-Rays   *page 2-4-10*

TABLE 2-4-8: Action to Take After a Veterinarian Evaluates the Dog's Eating Habits   *page 2-4-11*

Table 2-6-1: Determine Whether to Retire a Detector Dog When a Canine Officer Leaves the Position   *page 2-6-2*

TABLE 2-7-1: Implementation Process for a New Detector Dog Team   *page 2-7-2*

TABLE 2-7-2: Action to Take to Ensure Consistent Proficiency of a Detector Dog When a Canine Officer Is Away for Extended Periods   *page 2-7-7*

Table 3-2-1: Summary of Important Infectious Diseases   *page 3-2-6*

Table 3-2-2: Summary of External Parasites   *page 3-2-16*

Table 3-2-3: Summary of Internal Parasites   *page 3-2-17*

TABLE 3-3-1: Location of First Aid Related Topics in the Manual   *page 3-3-1*

TABLE 3-3-2: Signs of and First Aid for Bleeding   *page 3-3-3*

TABLE 3-3-3: How to Prevent Bloating   *page 3-3-4*

TABLE 3-3-4: Signs of and First Aid for Bloating   *page 3-3-4*

TABLE 3-3-5: Signs of and First Aid for Cold Injuries   *page 3-3-5*

TABLE 3-3-6: Signs of and First Aid for Foreign Objects in the Mouth   *page 3-3-5*

TABLE 3-3-7: Signs of and First Aid for Overheating (Hyperthermia)   *page 3-3-7*

Tables

TABLE 3-3-8: Signs of and First Aid for Shock   *page 3-3-11*

TABLE 3-4-1: Canine Vital Signs   *page 3-4-2*

TABLE 3-4-2: Daily Health Check   *page 3-4-4*

TABLE 3-4-3: Health Abnormality Checklist   *page 3-4-10*

TABLE 3-4-4: Symptom Checklist   *page 3-4-10*

TABLE 3-4-5: Determining Action to Take on Sudden Changes in Your Dog's Appetite   *page 3-4-12*

TABLE 3-4-6: Determining Action to Take on Coughing (Including Wheezing, Sneezing, and "Reverse Sneezing")   *page 3-4-15*

TABLE 3-4-7: Diarrhea and Bowel Movement Irregularities   *page 3-4-17*

TABLE 3-4-8: Determining Action to Take for Diarrhea and Bowel Movement Irregularities   *page 3-4-18*

TABLE 3-4-9: Determining Action to Take for Vomiting   *page 3-4-20*

TABLE E-1-1: Where to Check About Needed Quarantine or Health Requirements for Arriving Animals   *page E-1-2*

Figures

FIGURE 2-2-1: Detector Dog Assault Report   *page 2-2-3*

FIGURE 2-2-2: Sample of Standard Form 95, Claim for Damage, Injury, or Death   *page 2-2-6*

FIGURE 2-2-3: Detector Dog Aggression Report   *page 2-2-7*

FIGURE 2-3-1: A Checklist of Basic Kenneling Requirements   *page 2-3-3*

FIGURE 2-3-2: A Checklist of Sanitary Requirements   *page 2-3-4*

FIGURE 2-4-1: Model of a Limited Release Form and Sterilization Agreement   *page 2-4-6*

FIGURE 2-4-2: AVMA Positioning Criteria for X-rays   *page 2-4-10*

FIGURE 2-4-3: Model of a Final Release Form   *page 2-4-12*

FIGURE 2-5-1: Media Calls   *page 2-5-5*

FIGURE 2-6-1: Sample of Acknowledgment of Receipt, Agreement and Waiver of Liability   *page 2-6-4*

FIGURE 3-1-1: External Anatomy of a Beagle   *page 3-1-2*

FIGURE 3-1-2: Internal Anatomy of a Beagle—Skeletal Structure   *page 3-1-3*

FIGURE 3-1-3: Internal Anatomy of a Beagle—Internal Organs (male)   *page 3-1-4*

FIGURE 3-1-4: Internal Anatomy of a Beagle—Internal Organs (female)   *page 3-1-5*

FIGURE 3-2-1: Life Cycle of a Heartworm   *page 3-2-12*

FIGURE 3-2-2: Life Cycle of a Hookworm   *page 3-2-13*

FIGURE 3-2-3: Life Cycle of a Tapeworm   *page 3-2-15*

FIGURE 3-4-1: Placement of Hand Over the Dog's Muzzle   *page 3-4-23*

FIGURE 3-4-2: How to Administer Capsules or Tablets   *page 3-4-24*

FIGURE 3-4-3: Stroking the Dog's Throat to Facilitate Swallowing Medicine   *page 3-4-24*

FIGURE 3-4-4: Positioning of a Medicine Dropper or Syringe to Administer Liquids   *page 3-4-25*

FIGURE 3-4-5: How to Hold Dog's Head While Administering Eye Ointment   *page 3-4-26*

FIGURE 3-4-6: How to Apply Eye Ointment   *page 3-4-26*

FIGURE 3-4-7: Applying Medicine to Ears   *page 3-4-27*

FIGURE 3-4-8: Massaging Ears   *page 3-4-27*

Figures

FIGURE 3-5-1: Blood Supply of a Dog's Nail   *page 3-5-4*

FIGURE 3-5-2: Blood Supply Recedes as the Nail is Trimmed   *page 3-5-4*

FIGURE 3-5-3: Properly Trimmed Nail   *page 3-5-4*

FIGURE 3-5-4: Cleaning Ear   *page 3-5-5*

FIGURE 3-5-5: Massaging Ear   *page 3-5-5*

FIGURE 3-5-6: Allowing the Dog to Shake its Head   *page 3-5-5*

FIGURE 3-5-7: Wiping the Ear   *page 3-5-6*

FIGURE D-1-1: Collars Worn by Detector Dogs   *page D-1-2*

FIGURE D-1-2: Emergency Elizabethan Collar Made from a Plastic Planting Pot   *page D-1-3*

FIGURE D-1-3: Martingale Collar   *page D-1-3*

FIGURE D-1-4: Regular Nylon Collar   *page D-1-4*

FIGURE D-1-5: Slip Collar   *page D-1-5*

FIGURE D-1-6: Correct Way to Place a Slip Collar   *page D-1-5*

FIGURE D-1-7: Incorrect Way to Place a Slip Collar   *page D-1-6*

FIGURE D-1-8: Harness   *page D-1-6*

FIGURE D-1-9: Harness Placement   *page D-1-6*

FIGURE D-1-10: Wire Crates Used for Detector Dogs   *page D-1-7*

FIGURE D-1-11: Portable Kennels Used for Detector Dogs   *page D-1-8*

FIGURE D-1-12: First Aid Kit Supplied by NDDTC   *page D-1-9*

FIGURE D-1-13: Trauma Kit   *page D-1-12*

FIGURE D-1-14: A Variety of Grooming Tools   *page D-1-12*

FIGURE D-1-15: Leashes (Regular and Retractable)   *page D-1-14*

FIGURE D-1-16: Reward Pouch   *page D-1-15*

FIGURE G-1-1: Location of Anal Glands   *page G-1-1*

FIGURE H-1-1: Sample of a Canine Pest Identification Log   *page H-1-4*

FIGURE H-1-2: Agriculture Detector Dog Training Record with Keyed Areas   *page H-1-7*

FIGURE H-1-3: Example of a Completed Agriculture Detector Dog Training Record   *page H-1-8*

FIGURE H-1-4: Significant Incident Report   *page H-1-17*

FIGURE H-1-5: Significant Incident Report Continuation Sheet   *page H-1-18*

1 Introduction

National Detector Dog Manual

Contents

Purpose **page-1-1-1**
Scope **page-1-1-1**
   What the National Detector Dog Manual Does Not Cover **page-1-1-3**
Users **page-1-1-3**
Key Contacts **page-1-1-3**
Roles and Responsibilities **page-1-1-4**
History of USDA-APHIS Detector Dogs **page-1-1-4**
Vision and Activity Goals **page-1-1-5**
  Vision Statement **page-1-1-5**
  Activity Goals **page-1-1-5**
Related Documents **page-1-1-6**
Conventions **page-1-1-7**

Purpose

This manual has the following three purposes:

1. A reference guide for experienced Canine Officers to assist them in performing their duties.

2. A training tool for orienting new Canine Officers.

3. General information for secondary users, such as guidelines for supervisors and managers of Canine Officers, and Regional Canine Program Coordinators (RCPCs).

Scope

The National Detector Dog Manual covers background information, procedures, health care, and training related to detector dog activities. The procedures have a national focus to guide detector dog activities, and they are supplemental to general operational procedures in the Airport and Maritime Operations Manual (AMOM).

Introduction:
Scope

This manual is divided into five chapters:

- **Introduction**
- **Procedures**
- **Health Care**
- **Training**
- **Glossary**

Also included are appendixes and an index.

The ***Introduction*** chapter provides basic information about the manual. The information includes its purpose, scope, users, related documents, and a description of unfamiliar or unique symbols and highlighting that are used throughout the manual; the history of detector dogs in the United States Department of Agriculture (USDA), Animal and Plant Health Services (APHIS) and activity goals of detector dog activities in Plant Protection Quarantine (PPQ).

The ***Procedures*** chapter provides national-level guidelines for implementing and maintaining proficient detector dog teams at PPQ ports of entry as an alternative inspection technique. The sections cover the equipment needed, kenneling requirements, procuring detector dogs, how to manage incidents, public awareness, utilizing and retiring detector dogs. This chapter is tabbed as follows:

- **Equipment**
- **Incidents**
- **Kenneling Requirements**
- **Procuring Detector Dogs**
- **Public Awareness**
- **Retiring Detector Dogs**
- **Utilizing Detector Dogs**

The ***Health Care*** chapter provides basic facts and guidance for taking care of detector dogs. This chapter covers the external and internal anatomy of a dog, the common diseases and parasites of dogs, first aid and emergency care techniques, and general care of detector dogs. The chapter is tabbed as follows:

- **Anatomy**
- **Diseases and Parasites**
- **First Aid and Emergency Care**
- **General Care**
- **Grooming**

The ***Training*** chapter provides information about the training conducted for the detector dog team.

The ***Glossary*** defines specialized words, abbreviations and acronyms, training terms, and other difficult terms used to implement and manage detector dog activities.

The *Appendixes* list information that support the remaining content of the manual. The appendixes include history and trivia about beagles; lists of contacts within APHIS, roles and responsibilities that support detector dog activities, and equipment; guidelines for reporting and documenting results, for rating the weight of a dog, for shipping and daily transporting detector dogs, and for keeping the manual updated.

What the National Detector Dog Manual Does Not Cover

This manual does not cover:

- Local and regional policy and guidelines that should expand the national guidance provided in this manual.

- General, operational guidance provided by other PPQ manuals such as the following:

    - Operations from the AMOM

    - Monitoring from the Agricultural Quarantine Inspection Monitoring (AQIM) Handbook

    - Treatments from the PPQ Treatment Manual and the Animal Product Manual

Users

The primary users of this manual are PPQ Canine Officers. Secondary users include supervisors, Port Directors, RCPCs, the National Detector Dog Program Manager (NDDPM), Training Specialists, Animal Caretakers, State Plant Health Directors (SPHDs), State Operational Support Officers, Regional Directors, Regional Program Managers, headquarters staff, other Federal agencies, and foreign governments.

Key Contacts

PPQ detector dog teams are deployed at work locations across the country. The teams are supported by program managers at the regional and national levels, and program coordinators at the regional level. Refer to ***Appendix A*** for a directory of addresses, telephone numbers, and FAX numbers of the PPQ work locations that support detector dog teams.

Introduction:
Communications

Communications

All communication regarding the detector dog program must be channelled through the RCPC.

Roles and Responsibilities

PPQ's detector dog activities are managed within the regional structure by the RCPCs. At their assigned work location, the detector dog teams may be directed by a supervisor, manager, or port director. Supervisory and administrative support are provided through normal PPQ channels.

Those who support detector dog activities in PPQ hold the following positions:

- Canine Officers
- Co-workers
- Local managers (supervisors, port directors)
- Regional Canine Program Coordinators (RCPCs)
- Regional Program Managers
- National Detector Dog Instructors at the National Detector Dog Training Center (NDDTC)
- Animal Care Technicians
- Professional Development Center (PDC)
- National Detector Dog Program Manager (NDDPM)

Refer to **Appendix B** for examples of performance elements and for roles and responsibilities of these positions. What is listed in Appendix B is not all inclusive of the tasks performed by those who hold the positions.

History of USDA-APHIS Detector Dogs[1]

In 1984, USDA-APHIS began a detector dog program at Los Angeles International Airport with one detector dog team consisting of a beagle and a Canine Officer.

At first, APHIS tried a variety of dog breeds and worked with U.S. Customs to develop a detector dog program. As a result of this initial work, beagles were selected as the first detector dogs because of their acute sense of smell and their gentle nature with people. Refer to **Appendix C, *History of Detector Dog Programs***, for additional information about the beagle breed.

After selecting beagles as the Agency's first detector dogs, APHIS worked with the military at Lackland Air Force Base in Texas to train the first detector dog teams. The first class was held in 1986. Much of the early USDA detector dog training was modeled after methods used by the United States Air Force. Training methods have evolved since the inception of the program to include methods based on successful practical experience. Agency contributors to the creation of the detector dog program were the following: Douglas R. Ladner, PPQ Senior Staff Officer, and Mike Simon, Mel Robles, Cal Brannaka, and Hal Fingerman, all PPQ Canine Officers.

In 1987, APHIS opened three regional training centers and began training its own detector dog teams in 1988. The regional training centers were located in New York, Miami, and San Francisco, each staffed with one trainer who began conducting pilot classes in 1988. As the program grew, and training and support needs changed, a national training center was implemented created and the three regional training centers were consolidated. In October 1997, the National Detector Dog Training Center (NDDTC) officially opened.

Now detector dog teams are located at all major airports across the United States. Also, detector dog activities have been expanded to mail facilities, land border crossings, and ports that handle cargo.

Vision and Activity Goals

Vision Statement
Deploy detector dogs in all areas where they can be most effectively utilized and integrated into the operations of APHIS and PPQ.

Activity Goals
1. Establish proficiency levels that all detector dog teams must maintain to be effective in protecting American agriculture.

---

1 Previously known as the "Beagle Brigade"

Introduction:
Related Documents

2. Successfully integrate the detector dog activity into the agricultural quarantine inspection (AQI) operation in:
   A. Baggage clearance at airports, maritime ports, ships, and military facilities
   B. International mail and small parcel clearance (eg., USPS, DHL and FedEx)
   C. Bulk and containerized cargo clearance at airports and maritime ports
   D. Vehicle, cargo, and baggage clearance at land border crossings
   E. Smuggling interdiction in all venues
   F. Hawaii domestic mail program
3. Deliver excellent and timely training that will support the local, regional, and national AQI and related programs.
4. Provide training to supervisors and port directors so they can effectively manage detector dog activities.
5. Explore activities in addition to traditional AQI in which detector dogs may effectively be used.

Related Documents

- Detector Dog Program Training Manual
- Legislative and Public Affairs (LPA) Pamphlets
- 9CFR Parts 1, 2, and 3, Humane Treatment of Dogs and Cats; Temperature Requirements
- 9CFR Chapter 1
- AMOM and other related import manuals

- Animal Welfare Act
- Traveling With Your Pet, Miscellaneous Publication No. 1536
- Beagle Brigade web page at the following address: http://www.aphis.usda.gov/travel
- Bill H.R. 2559 (Barney Bill)

Conventions

Bullets

Bulleted lists indicate that there is no hierarchical order in the information being listed. Bullets and sub-bullets look like this in the manual:

- Collars
  - Slip
  - Leather
  - Nylon

Caution

A caution advisory indicates that people or dogs could possibly be endangered or slightly hurt. Compare to "warning." A caution advisory looks like the following throughout the manual:

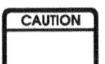

Chapter

This manual contains the following five chapters: **Introduction**, **Procedures**, **Health Care**, **Training**, and **Glossary**.

Chapter Table of Contents

Almost every section in each chapter has a table of contents. Those are at the beginning of the chapter or section and facilitate finding information within a section. The table of contents lists the heading titles within each section.

Control Data

Information placed at the top and bottom of each page helps users navigate the manual and its updates. At the top of the page is the chapter, section, and first-level heading. At the bottom of the page is the month, year, and version of the manual, its title, page number, and the unit responsible for the content.

Heading Levels

Within each section there are three headings. The first heading is within a horizontal line and followed by the title that continues across both the left and right columns. The second heading is in the right-hand column with the text beginning below it. The third heading is in the left-hand column and is used to easily scan topics.

Highlighting Tables, Figures, Sections

When tables, figures, or other sections are referenced in the body of the manual, they are emphasized *in italic print*. For example, refer to the *Glossary* for a definition of primary residence.

Important Note

Helpful hint(s) or other information that assists the user look like the following throughout the manual:

Must

When must is used in this manual, it means mandatory by policy.

Notice

Notices indicate a dangerous situation is possible in which goods might be damaged. A notice looks like the following throughout the manual:

Numbering Scheme

A three-level numbering scheme is used in this manual for pages, tables, and figures. The first number represents the chapter. The second number represents the section. The third number represents the page, table, or figure. This numbering scheme allows for easier updating and adding of pages without having to reprint an entire chapter. Dashes are used in page numbering to differentiate page numbers from decimal numbers.

Introduction:
Conventions

Section
Two of the larger chapters in this manual are broken down into sections. Each section begins on a new, right-hand page. Sections are identified by the second number in the numbering scheme used for this manual. The **Procedures** and **Health Care** manual chapters have sections,

Tab Colors
Only two tabs are colored. The section titled **Incidents** has a red tab, and the section titled **First Aid and Emergency Care** has a green tab. The users are able to quickly find guidance when confronted with a major incident such as a dog aggression incident, or when administering first aid and emergency care to a detector dog.

Warning
A warning advisory indicates that people or dogs could possibly be hurt or killed. Compare to "Caution." A warning advisory looks like the following throughout the manual:

Introduction:
Conventions

Procedures

Equipment

Contents

Introduction **page-2-1-1**
NDDTC Supplied Equipment and Supplies **page-2-1-1**
Field Supplied Equipment and Supplies **page-2-1-2**

Introduction

This section contains a list of the recommended equipment needed to support a detector dog team at a work location. The equipment should be acquired as soon as possible. The list is divided between the equipment supplied by NDDTC when the Canine Officer attends Basic Canine Officer Training (BCOT), and the equipment supplied by the field at the work location. **Canine Officers are not authorized to use any additional equipment without approval from the RCPC.**

NDDTC Supplied Equipment and Supplies

The following are supplied by NDDTC:

- **Collars** (see ***Appendix D*** for more information about collars including Elizabethan and Martingale collars)
  - Slip
  - Nylon
- **Leashes (Regular and Retractable)**, (see ***Appendix D*** for more information)
- **Jackets** (see ***Appendix D*** for more information)
- **Grooming Kit** (see ***Appendix D*** for more information)
- **Crates/Portable Kennels** (see ***Appendix D*** for more information)
- **Reward Pouch** (see ***Appendix D*** for more information)
- Water bucket, 2-quart stainless steel
- Detector dog **First Aid Kit** (see ***Appendix D*** for more information)
- National Detector Dog Manual

Field Supplied Equipment and Supplies

The following are supplied by the field and should be acquired as soon as possible:

- Vehicle (see **Appendix E** for **Shipping and Daily Transporting Detector Dogs**. See **Appendix D** for a list of safety requirements under **Vehicles**.)
- Collars and leashes, additional
- Crate pads (see **Appendix D** for more information under **Crates/Portable Kennels**; see **Appendix E** for **Shipping Detector Dogs**.)
- Additional crates—one for vehicle (wire cage), one for the office, and one for each inspectional site, such as post office or airport
- Additional water buckets, 2-quart stainless steel
- Flea comb
- Pooper scooper
- Styptic powder to help stop bleeding caused by trimming nails and minor cuts
- Communication device for safety, such as car phone or 2-way radio (as directed by the region or work location)
- Credit card for purchasing supplies—minimum of one at a work location for the detector dog team
- **Refrigerators** for storing training aids designated for the detector dog team (see **Appendix D** for more information)
- Air-tight containers for storing training aids in the refrigerator(s)
- Target items, i.e., meats (pork and beef), fruits and vegetables, soil, plants, seeds, etc.

Important

The amount and type of target and nontarget items will vary among work locations and should be determined locally by Canine Officers. As a guide, the more the better; you need a sufficient amount to create realistic passenger baggage scenarios for training exercises, including boxes.

- Nontarget items typical of flights encountered at the work location, i.e., cosmetics, toiletries, candy, chips, coffee, cheeses, fish, bread, chocolate, etc.
- Filler items, i.e., clothes, shoes, etc.
- Cardboard boxes and paper bags (see **Appendix D** under **Suitcases, Boxes, and Contents** for more information)

Procedures: Equipment
Field Supplied Equipment and Supplies

- Suitcases and handbags—minimum of 50 hard and soft cases in a variety of sizes; including 50 percent handbags, backpacks, purses; excluding target cases (see **Appendix D** under **Suitcases, Boxes, and Contents** for more information). Additionally, maintain 25 boxes.
- Access to:
  - Office equipment for preparing and distributing reports, such as a computer, a FAX machine, an assigned e-mail address for Lotus Notes®
  - A storage area for training equipment, such as suitcases, boxes, refrigerators, target and nontarget items, filler items
  - A storage area for crates
  - Digital camera
  - Video camera (camcorder)

Procedures: Equipment
Field Supplied Equipment and Supplies

2 Procedures

Incidents

National Detector Dog Manual

Contents

Introduction **page-2-2-1**
Assaults Toward a Detector Dog   **page-2-2-1**
Dog Aggression   **page-2-2-4**
Emergency Response Plan   **page-2-2-8**
Injury and Sudden Illness   **page-2-2-8**
Incident Contacts   **page-2-2-9**

Introduction

Report incidents involving assaults toward a detector dog, dog aggression and/or situations involving a canine team (example: passenger assaults a canine officer) to the Commissioner's Situation room. Contact local CBP management and the RCPC, following the directions in this manual for the particular situation. Refer to **CBP Directive No. 3340-025B: Commissioner's Situation Room Reporting** in **Appendix H** for further guidance.

Assaults Toward a Detector Dog

If your detector dog is assaulted or interfered with, consult with your Regional Canine Program Coordinator (RCPC) and immediately notify your local port manager.

Important

Under authority of Sec. 501, Title V of Bill H.R. 2559 (commonly known as the Barney Bill), anyone who intentionally harms or interferes with a USDA detector dog may face a fine of up to $10,000. See **Appendix J** for the legislative authority. Refer to the following steps if your dog is assaulted. If the injury or illness is life threatening, contact your supervisor as soon as possible. If the incident is critical but not life threatening, contact your supervisor before taking the dog to the veterinarian. The Canine Officer is responsible for contacting the RCPC.

1. Stabilize the dog using the procedures for "Injury and Sudden Illness" on **page-2-2-8**.
2. Get a witness statement from anyone who saw the attack. This includes the PPQ Canine Officer. The statement should include who, what, when, where, and why. Fill out a detector dog assault report (see **Figure 2-2-1**).
3. If the attacking passenger is cooperative, get a statement from him/her. Ask local security for assistance in talking to the attacking passenger, if necessary.

Procedures: Incidents
Assaults Toward a Detector Dog

4. Get identifying information of the attacking passenger, including the name, address, time of day, and airline on which he or she was flying. Get a copy of the passport and declaration.
5. Photograph the dog, if it has visible injuries, before the visit to the veterinarian, if possible.
6. Take the dog to a veterinarian immediately for examination. Be sure to get the veterinarian's statement.
7. Contact your RCPC immediately.
8. Contact your local Investigative Enforcement Services (IES).

Do **not** write a civil penalty, as legal action will be taken based on the IES investigation.

Procedures: Incidents
Assaults Toward a Detector Dog

Detector Dog Assault Report

Name _____   Canine _____
Duty Location _____   Phone _____
Date/Time of Statement _____   Date/Time of Incident _____

Please answer the following questions regarding the incident:

1. Did you witness the incident?   Yes _____   No _____

2. Was the detector dog injured as a result of the incident?   Yes _____   No _____

If yes, describe the injuries in detail _____
_____
_____

3. Was the aggressor injured in any way by the detector dog?   Yes _____   No _____

   ◆ If yes, complete a Detector Dog Aggression Report.

4. Were there other witnesses to the incident?   Yes _____   No _____

If yes, please list the witnesses' names and contact numbers on a separate piece of paper, attached to this report. If possible, have them fill out a separate Detector Dog Assault Report and attach to this form.

   ◆ Get identifying information of attacking passenger, including the name, address, time of day, and airline he or she was flying on. Photocopy information concerning the attacker: Customs card, passport, driver's license. Attach information to this report.

   ◆ Follow all instructions on assaults from the National Detector Dog Manual.

Describe your observation of the assault in detail (attach sheet if needed).

Attach any photographs.

FIGURE 2-2-1: Detector Dog Assault Report

Dog Aggression

It is important to collect a detector dog aggression report from each individual who witnessed the incident in its entirety. If any person (including a Canine Officer, a kennel worker, or a passenger) is allegedly bitten by a detector dog or if the detector dog shows any aggression toward a person, then do the following:

1. If the dog behaves aggressively, immediately remove it from the work environment and contact local port management.

2. Secure the dog in a crate until you can take it to the veterinarian for a physical exam. The medical evaluation should be conducted within 48 hours and should include tests for hormonal balance, structural or soft tissue pain or discomfort, a neurological consultation, urine metabolite screening (especially for excessive levels of glutamine, associated with neuronal death) and allergies.

3. If someone is bitten or is allegedly bitten, take the person to a quiet place, such as an office. Call emergency medical service and administer first aid, if necessary. If there is bleeding, use precautions.

4. Get the following information about the person who was allegedly bitten:

    A. Name

    B. Address

    C. Other pertinent information—medications used, permanent residence or temporary residence while in the United States, if a passenger. Make a copy of the passport, customs declaration, and driver's license.

    D. If the person refuses emergency medical service, make note of the refusal. Try to get the person's signature on a statement of refusal of emergency medical service (SF 95).

    E. Have the individual and all witnesses complete the detector dog aggression report.

    F. Photograph the injury if possible.

    G. If the aggression incident occurred in the Federal Inspection Service (FIS) area, note it on the passenger's declaration card. Make a copy of the card.

5. If the person goes to a hospital, notify the nearest Office of General Counsel (OGC) at http://dc-directory.hqnet.usda.gov/phone.php. Each work location should have the telephone number of the nearest OGC available in case it is needed. Record the number at the end of this section.

6. Direct the victim to complete a Standard Form 95, Claim for Damage, Injury, or Death (Standard Form 95A is a Spanish version). Refer to **Figure 2-2-2** for a sample of the form. Direct the victim to return the form to the local PPQ office or to the following address:

   USDA-APHIS-ABS
   Accounting and Property Services
   100 N. Sixth Street, 5th Floor
   Minneapolis, MN 55403

   For further information about procedures for tort claims, see Departmental Regulation 2510-1, Claims Against the United States, dated July 20, 1992. Work locations should have this regulation on file.

7. Write a detailed detector dog aggression report as soon as possible. Each work location decides the protocol for notifying management after duty hours. Refer to **Figure 2-2-3** for a sample of the form.

8. Submit the complete packet to the RCPC **within 72 hrs. of the incident.** Await further instructions regarding the detector dog.

9. Do not allow the detector dog back into service until notified by your RCPC. The incident will have to be investigated thoroughly by your RCPC. The RCPC will inform the RPM, who will inform the NDDPM of the aggressive incident or bite.

Procedures: Incidents
Dog Aggression

CLAIM FOR DAMAGE, INJURY, OR DEATH	**INSTRUCTIONS:** Please read carefully the instructions on the reverse side and supply information requested on both sides of the form. Use additional sheet(s) if necessary. See reverse side for additional instructions.	FORM APPROVED OMB NO. 1105-0008
1. Submit To Appropriate Federal Agency:	2. Name, Address of claimant and claimant's personal representative, if any. *(See instructions on reverse.) (Number, street, city, State and Zip Code)*	

3. TYPE OF EMPLOYMENT ☐ MILITARY ☐ CIVILIAN	4. DATE OF BIRTH	5. MARITAL STATUS	6. DATE AND DAY OF ACCIDENT	7. TIME *(A.M. or P.M.)*

8. Basis of Claim *(State in detail the known facts and circumstances attending the damage, injury, or death, identifying persons and property involved, the place of occurrence and the cause thereof) (Use additional pages if necessary.)*

9. **PROPERTY DAMAGE**

NAME AND ADDRESS OF OWNER, IF OTHER THAN CLAIMANT *(Number, street, city, State, and Zip Code)*

BRIEFLY DESCRIBE THE PROPERTY, NATURE AND EXTENT OF DAMAGE AND THE LOCATION WHERE PROPERTY MAY BE INSPECTED. *(See instructions on reverse side.)*

10. **PERSONAL INJURY/WRONGFUL DEATH**

STATE NATURE AND EXTENT OF EACH INJURY OR CAUSE OF DEATH, WHICH FORMS THE BASIS OF THE CLAIM. IF OTHER THAN CLAIMANT, STATE NAME OF INJURED PERSON OR DECEDENT.

11. **WITNESSES**

NAME	ADDRESS *(Number, street, city, State, and Zip Code)*

12. *(See instructions on reverse)*	**AMOUNT OF CLAIM** *(In dollars)*			
12a. PROPERTY DAMAGE	12b. PERSONAL INJURY	12c. WRONGFUL DEATH	12d. TOTAL *(Failure to specify may cause forfeiture of your rights.)*	

I CERTIFY THAT THE AMOUNT OF CLAIM COVERS ONLY DAMAGES AND INJURIES CAUSED BY THE ACCIDENT ABOVE AND AGREE TO ACCEPT SAID AMOUNT IN FULL SATISFACTION AND FINAL SETTLEMENT OF THIS CLAIM.

13a. SIGNATURE OF CLAIMANT *(See instructions on reverse side.)*	13b. Phone number of signatory	14. DATE OF CLAIM

CIVIL PENALTY FOR PRESENTING FRAUDULENT CLAIM	CRIMINAL PENALTY FOR PRESENTING FRAUDULENT CLAIM OR MAKING FALSE STATEMENTS
The claimant shall forfeit and pay to the United States the sum of $2,000 plus double the amount of damages sustained by the United States. *(See 31 U.S.C. 3729.)*	Fine of not more than $10,000 or imprisonment for not more than 5 years or both. *(See 18 U.S.C. 287, 1001.)*

95-109
Previous editions not usable.
Designed using Perform Pro, WHS/DIOR, Jun 98

NSN 7540-00-634-4046

STANDARD FORM 95 (Rev. 7-85) (EG)
PRESCRIBED BY DEPT. OF JUSTICE
28 CFR 14.2

FIGURE 2-2-2: Sample of Standard Form 95, Claim for Damage, Injury, or Death

Procedures: Incidents
Dog Aggression

Detector Dog Aggression Report

Name _____   Canine _____
Duty Location _____   Phone _____
Date/Time of Statement _____   Date/Time of Incident _____

Please answer the following questions regarding the incident:

1. Did you witness the incident?   Yes _____   No _____

2. What type of incident was it?

Any form of aggression towards the detector dog _____

Re-directed aggression _____

Medical reason (i.e. seizure) _____

Other (i.e. food grabbing) _____

3. Was there a wound as a result of the incident?   Yes _____   No _____

    If yes, was the skin broken?   Yes _____   No _____

    If yes, was medical attention required?   Yes _____   No _____

Describe the injuries in detail _____
_____
_____

4. Was the dog assaulted as a result of this incident?   Yes _____   No _____
   ◆ If yes, complete a Detector Dog Assault Report.

5. Were there other witnesses to the incident?   Yes _____   No _____

If yes, please list the witnesses' names and contact numbers on a separate piece of paper, attached to this report. If possible, have them fill out a separate Detector Dog Aggression Report and attach to this form.

Describe your observation of the incident in detail (attach sheet if needed).

Attach any photographs.

FIGURE 2-2-3: Detector Dog Aggression Report

Emergency Response Plan

Every PPQ work location should have on file an emergency response plan. This plan directs local managers when confronted with civil disturbances and natural disasters, such as flood, fire, hurricane, tornado, earthquake, or inclement weather. At work locations where there is a detector dog team, the emergency response plan should include plans to secure the safety of the detector dog.

The plan should address the following issues for securing the detector dog. This list is not inclusive.

- Who is responsible for implementing the plan
- Emergency veterinary care
- Alternative kenneling
- Methods of providing basic needs of the detector dog, such as food, water, and exercise
- Installation of smoke detectors in detector dogs' working environment

Injury and Sudden Illness

The Canine Officer is responsible for determining when the detector dog needs medical care and must ensure the dog's needs are met when it is injured or ill.

1. Stabilize the dog.
2. Administer first aid or emergency care. Refer to **First Aid and Emergency Care** information behind the green tab.
3. If the first aid indicates to immediately take the dog to the veterinarian, then do so.

Important

If the injury or illness is life threatening, contact your supervisor as soon as possible. If the incident is critical but not life threatening, contact your supervisor before taking the dog to the veterinarian. The Canine Officer is responsible for contacting the RCPC.

4. Require a written release from the veterinarian stating that the dog is able to return to work with no restriction.

Detector dogs that have been injured or ill and under a veterinarian's care will not return to work until the release is provided. Therefore, Canine Officers need to keep their supervisors informed since they are assigning and directing work activities. The Canine Officer will send a copy of the release to the RCPC.

5. Get a copy of the veterinary bill.

Incident Contacts

The remainder of this section is for you to record local contacts you wish to have in the manual in case of an incident.

Veterinarian:

___

___

Local emergency or veterinary service:

___

___

National Animal Poison Control Center:
1-900-680-0000
1-800-548-2423
Local:

___

___

Police:

___

___

Office of General Counsel:

___

___

Other local contacts:

___

___

Procedures: Incidents
Incident Contacts

Procedures

Kenneling Requirements

Contents

Introduction **page-2-3-1**
Initial Selection of a Kennel **page-2-3-1**
Basic Kenneling Requirements **page-2-3-2**
Sanitary Requirements **page-2-3-4**
Monitoring Kenneling Services **page-2-3-5**

Introduction

Kenneling detector dogs is an important responsibility of Canine Officers, in conjunction with port directors and RCPCs.

The kennel environment has a tremendous influence on a detector dog's mental and physical well being. Therefore, detector dogs should be kenneled at facilities that maintain high standards of cleanliness and security. Ideally, select a kennel with indoor/outdoor runs and an exercise area.

Initial Selection of a Kennel

If you work at a location that does not already have a detector dog team in place, then a suitable kennel must be contracted for services.

The following is a list of possible places or people to contact for potential kennels:

- U.S. Customs—where do they kennel their detector dogs? Possibly establish inter-agency agreement to share kennel facilities
- Personal recommendations of coworkers—where do they board their pets?
- Society for the Prevention of Cruelty to Animals—may have a list of available kennels
- Kennel associations, animal shelters, humane organizations
- Yellow pages
- Veterinarians
- RCPCs

Procedures: Kenneling Requirements
Basic Kenneling Requirements

The administrative tasks associated with selecting a kennel and annually renewing this service varies among regions and work locations. Therefore, consult with the port director and the RCPC when selecting a kennel. Administrative tasks may include the following:

- Preparing contract specifications
- Negotiating price
- Preparing a Procurement Request, AD Form 700
- Budgeting for services
- Renewing contract services

Use the basic kenneling and sanitization requirements listed in this section as a checklist when visiting a potential kennel to evaluate their standards of cleanliness and security. (See *Basic Kenneling Requirements* and *Sanitary Requirements*.)

Basic Kenneling Requirements

When visiting kennels, pay attention to the general appearance, atmosphere, and smell of the facility. Owners should be willing to show you around.

Figure 2-3-1 is a checklist of general requirements to look for when selecting a suitable kennel. Given that the final selection may be a compromise or conditional, the requirements are divided between those that are a "must" and those that the kennel "should have." Lacking "should have" requirements should not exclude a kennel.

Procedures: Kenneling Requirements
Basic Kenneling Requirements

A Kennel Must:

- ❏ Separate healthy dogs from nonhealthy dogs by maintaining an area to quarantine nonhealthy dogs, and by taking the necessary precautions.
- ❏ Maintain a properly ventilated facility (12-15 air exchanges per hour recommended).
- ❏ Provide emergency care, or will transport dog to an emergency clinic.
- ❏ Heat and cool as dictated by climate to maintain 60–80 °F in the facility year round.
- ❏ Maintain feeding and medication schedules as directed by the Canine Officer.
- ❏ Provide in-house grooming service or an area where the Canine Officer can groom the dog.
- ❏ Provide 24-hour access.
- ❏ Provide timely and accurate communication to Canine Officers or other Agency personnel about the condition of the detector dog.
- ❏ Meet minimum requirements of the Animal Welfare Act for primary residence. (See the **Glossary** for the definition of **Primary residence**.)
- ❏ Maintain well-kept grounds free of tall grasses, overgrown shrubbery, and fallen leaves and plant debris.
- ❏ Provide Runs That:
  - ❏ Are escape proof (i.e., fence on top, secure latch).
  - ❏ Have concrete floors.
  - ❏ Have good drainage, no puddles.
  - ❏ Have a solid, impervious barrier preventing physical contact with other dogs to prevent cross-contamination.
- ❏ Have Good Security That:
  - ❏ Can be accessed only by a key after hours.
  - ❏ Prevents dogs from digging out or jumping or climbing over fences.
- ❏ Have a Clean and Sanitary Facility That:
  - ❏ Smells and appears clean.
  - ❏ Stores food in a dry, clean place.
  - ❏ Is free of rodent droppings.
  - ❏ Is free of poison baits or toxic material accessible to animals.
  - ❏ Is cleaned and sanitized daily. Sanitizing agents must be approved. (See *Figure 2-3-2 on page-2-3-4* for sanitation requirements.)

A Kennel Should Have:

- ❏ Detector dogs segregated from the general population, when possible.
- ❏ Detector dogs in adjacent kennels or in the same area.
- ❏ Locations as close to work as practical.
- ❏ Indoor and outdoor runs.
- ❏ Parking accessible to government-owned vehicles and privately-owned vehicles.
- ❏ An exercise area to allow the detector dog to run around in a safe, enclosed area.
- ❏ Smoke detectors.

FIGURE 2-3-1: A Checklist of Basic Kenneling Requirements

Sanitary Requirements

Cleanliness of kennels is an important factor for good health of detector dogs. Therefore, sanitary requirements must be enforced in and around the kennel. Sanitation is one of the main measures of disease prevention and control. Every Canine Officer must be concerned about a disease existing in one dog that might be passed on to the others.

There are many specific ways to keep a good level of sanitation in a kennel. Good sanitation can be maintained through a cooperative effort between Canine Officers and kennel personnel. Use *Figure 2-3-2* as a checklist to follow when evaluating sanitary requirements of a kennel.

The kitchen or food preparation area:

☐ Keep as clean as possible.

☐ Wash hands before preparing food.

☐ Clean utensils immediately after preparing food.

☐ If canned foods are being fed for a special diet, clean the can opener after each use.

☐ Maintain disinfectant procedures in the food preparation area.

Stools are a common source of infection:

☐ Remove from the runs as often as necessary.

☐ The method of disposing of stools depends on local conditions and the type of sewage system present.

☐ If stools must be carried from the area in cans, the cans must be cleaned and disinfected after each use.

In the kennels:

☐ Keep sanitary.

☐ Maintain in a good state of repair.

☐ Clean runs thoroughly on a daily basis.

☐ Disinfect periodically by using products suggested by your local veterinarian: Rocal®, Parvasol®, bleach (5.25 percent solution available chlorine).

   CAUTION: Before using any disinfectant, check the label to ensure proper use against dog-related diseases and viruses.

Around the kennels:

☐ Keep free of refuse and garbage that could attract rats and insects.

☐ Use mosquito control measures in areas as needed (i.e., netting or other measures).

☐ Use disinfectants and their application by following the label requirements. Disinfectants suggested by PPQ's VMO: Rocal®, Parvasol®, bleach (5.25 percent solution available chlorine).

FIGURE 2-3-2: A Checklist of Sanitary Requirements

Monitoring Kenneling Services

Once a facility is providing PPQ with kenneling service, Canine Officers are responsible for ensuring that all basic kenneling and sanitary requirements continue to be met.

Ultimately, supervisors are responsible for ensuring that a kennel continues to meet the basic kenneling and sanitary requirements and that the Canine Officer is satisfactorily monitoring this compliance.

Supervisors should visit the kennel at least twice a year.

Procedures: Kenneling Requirements
Monitoring Kenneling Services

Procedures

Procuring Detector Dogs

Contents

Introduction   **page 2-4-1**
Procuring Requirements   **page 2-4-2**
Sources   **page 2-4-2**
Summary of Procurement Process   **page 2-4-3**
Conduct Telephone Interview   **page 2-4-4**
Complete Limited Release Form and Sterilization Agreement   **page 2-4-5**
Conduct Initial Screening   **page 2-4-7**
Evaluate Temperament   **page 2-4-7**
Evaluate Health   **page 2-4-8**
  General Exam   **page 2-4-8**
  Heart Worm Test   **page 2-4-9**
  Blood Test   **page 2-4-9**
  X-rays   **page 2-4-9**
  Eating Habits   **page 2-4-11**
Complete Final Release Form   **page 2-4-11**
Complete Tracking Record and Feedback Worksheet   **page 2-4-13**
Complete Airline Flight Tracking Worksheet   **page 2-4-13**

Introduction

Procuring dogs is necessary to ensure that adequate numbers of high-quality detector dog candidates are available to the National Detector Dog Training Center (NDDTC) where they are prepared for detection work at international airports, border crossings, mail facilities, and cargo facilities throughout the United States. The goal of the Agency's detector dog program is to procure dogs that possess the following characteristics:

- Self-confidence
- Soundness
- High food drive
- Sociability
- Adaptability

The intent of this section is to familiarize Regional Canine Program Coordinators (RCPCs) and regionally selected Canine Officers to initially screen potential detector dogs, and local managers with the detector dog procurement process. RCPCs are responsible for ensuring that the appropriate health tests are administered and that proper shipping procedures are arranged prior to sending dogs to the NDDTC.

Procedures: Procuring Detector Dogs
Procuring Requirements

Following is a list of the documents that facilitate the procurement process. Each is explained in this section along with a sample that may be used by those designated to perform the steps of the process.

- Telephone Interview Worksheet
- Limited Release Form and Sterilization Agreement
- Initial Screening Process Worksheet
- Temperament Evaluation Worksheet
- Health Evaluation Protocol Worksheet
- Final Release Form
- Airline Flight Tracking Worksheet
- Tracking Record and Feedback Worksheet

Procuring Requirements

Canine Officers must meet the following requirements before being considered to procure dogs for the program:

1. Have at least three years with the Detector Dog Program
2. Have a fully successful evaluation
3. Maintain a detector dog team proficiency above 80 percent
4. Pass annual validation if given
5. Have permission of their managers
6. Be recommended by their RCPC
7. Be trained by RCPC to procure detector dogs
8. Be selected by RPM and RCPC

RCPCs will review Canine Officers' procurement lists annually. See **Appendix K** for a request to procure canines form.

Sources

Following is a list of suggested sources for seeking detector dogs. Typically, however, the procurement process begins when private individuals, rescue organizations, or shelters contact a Canine Officer or field work location.

- Local newspapers
- Internet (e.g., http://www.petfinders.org/)
- Breeders
- Breed rescue groups
- Shelters

Summary of Procurement Process

The procurement process is a sequence of steps that will ensure that the most qualified dogs are sent to the NDDTC for training and final preparation to become detector dogs. If a dog fails any part of the temperament evaluation or health screening, it cannot be accepted by the NDDTC. The steps of the process are summarized below, followed by their details in **Table 2-4-1**.

1. Interview the owners of the dog using the Telephone Interview Worksheet.
2. Contact an RCPC with the results of the interview to receive permission to proceed with an initial screening.
3. Complete a Limited Release Form and Sterilization Agreement if it is necessary to remove the dog from the owner's property to initially evaluate it.
4. Conduct an initial evaluation of the dog using the Initial Screening Process Worksheet.
5. If the dog passes the initial evaluation, complete a Limited Release Form and Sterilization Agreement so it can be removed from the owner's property to evaluate its temperament.
6. Evaluate the dog's temperament at a public location using the Temperament Evaluation Worksheet.
7. If the dog passes the temperament evaluation, contact the NDDTC to ensure that the dog can be accepted.
8. Schedule a veterinary appointment to evaluate the dog's health following the Health Evaluation Protocol Worksheet.
9. If the dog passes the health evaluation, complete a Final Release Form.
10. If you wish to monitor the status of the dog's evaluation and training at the NDDTC, complete a Tracking Record and Feedback Worksheet.
11. Ship the dog to the NDDTC using the Airline Flight Tracking Worksheet. Contact NDDTC prior to shipping.

Procedures: Procuring Detector Dogs
Conduct Telephone Interview

TABLE 2-4-1: Steps of the Procurement Process

Step:	Further Action to Take:	Document Step Using:
Interview the owner(s) of the dog	None	Telephone Interview Worksheet
Contact an RCPC with the results of the interview	Receive permission to proceed with an initial screening	No documentation required
Complete a limited release form if it is necessary to remove the dog from the owner's property to initially evaluate it	None	Limited Release Form and Sterilization Agreement
Conduct an initial evaluation of the dog	None	Initial Screening Process Worksheet
If the dog passes the initial evaluation	Complete a Limited Release Form so it can be removed from the owner's property to evaluate the dog's temperament	Limited Release Form and Sterilization Agreement
Evaluate the dog's temperament at a public location	None	Temperament Evaluation Worksheet
If the dog passes the temperament evaluation	Contact the NDDTC to ensure that the dog can be accepted	No documentation required
Schedule to have a veterinarian evaluate the dog's health	None	Health Evaluation Protocol Worksheet
If the dog passes the health evaluation	Complete a Final Release Form	Final Release Form
If you wish to monitor the status of the dog's evaluation and training at the NDDTC	Complete a Tracking Record and Feedback Worksheet	Tracking Record and Feedback Worksheet
Ship the dog to the NDDTC	Contact NDDTC prior to shipping	Airline Flight Tracking Worksheet

Conduct Telephone Interview

The telephone interview is used to determine whether to continue evaluating the dog as a potential candidate. The officers procuring detector dogs in the field should coordinate their efforts with their RCPCs and the NDDTC.

Use the Telephone Interview Worksheet in **Appendix K**. The worksheet has been printed so it can be removed, photocopied, and reused.

If the dog is **not** screened out in the telephone interview, arrange a time when the dog's owner(s) can be contacted for further evaluation.

Important

Contact an RCPC with the results of the telephone interview to receive permission to proceed with an initial screening.

Complete Limited Release Form and Sterilization Agreement

If it is necessary to remove the dog from the owner's property to initially evaluate it, complete a Limited Release Form and Sterilization Agreement. Refer to *Figure 2-4-1* for a model of the form and to *Appendix K* for a form you can copy.

Procedures: Procuring Detector Dogs
Complete Limited Release Form and Sterilization Agreement

Subject: Limited Release Form and Sterilization Agreement

I _____ do hereby give permission to
      (Owner)

_____ of the U.S. Department of Agriculture
(USDA Representative)

to take _____, _____
    (Name of Dog)   (Breed of Dog)

off my property for the sole purpose of temperament testing and health screening. Health screening will be done at no cost to me, the dog's owner. It is my understanding that I am **not** relinquishing legal claim or ownership at this time. I understand that if the dog is **not** accepted into the USDA Detector Dog Program, the dog will be returned to me at my expense. However, if the dog is accepted into the Program, the USDA assumes the responsibility to have

_____ spayed or neutered. At that time, a final
(Name of Dog)

release statement will be signed relinquishing my legal claim to the dog.

_____   _____
USDA Representative   Owner/Agent

_____
PPQ Work Location

Date: _____   Date: _____

FIGURE 2-4-1: Model of a Limited Release Form and Sterilization Agreement

Procedures: Procuring Detector Dogs
Conduct Initial Screening

Conduct Initial Screening

If the dog is **not** screened out in the telephone interview and permission is received from the RCPC and the NDDTC to continue screening the dog, the next step is to conduct an initial evaluation of the dog. The purpose of this screening is to determine if the dog initially meets the criteria as a potential candidate for the Agency's detector dog program. The initial screening covers the following areas:

- Food drive level
- Sociability
- Intelligence and ability to be trained
- Physical soundness
- Anxiety level

Use the Initial Screening Process Worksheet to evaluate the dog. The worksheet is located in **Appendix K** and has been printed so it can be removed, photocopied, and reused.

TABLE 2-4-2: Action to Take After Completing the Initial Screening

If the result of the initial screening is that the dog:	Then:
Passes	1. Complete a Limited Release Form and Sterilization Agreement, so the dog can be removed from the owner's property to evaluate its temperament
	2. Continue with the procurement process by evaluating the dog's temperament at a public place such as an airport
	NOTE: Consult with the RCPC on the results of the initial screening before continuing with the temperament evaluation
Fails	Arrange for the return of the dog to the owners at the owner's expense

Evaluate Temperament

The formal evaluation of the dog's temperament is ideally conducted at an airport, cargo facility, or border crossing for which the dog is being procured.

Use a Temperament Evaluation Worksheet to evaluate the dog's reaction to various stimuli and situations and to record a general impression of the dog's potential performance. The worksheet is located in **Appendix K** and has been printed so it can be removed, photocopied, and reused.

Procedures: Procuring Detector Dogs
Evaluate Health

TABLE 2-4-3: Action to Take After Completing the Temperament Evaluation

If the result of the temperament evaluation is that the dog:	Then:
Passes demonstrating a high food drive and confidence and stability in strange surroundings (Sum of mean ratings must total 18 or above)	1. Contact the RCPC with the results of the temperament test 2. Consult with the RCPC and the NDDTC to receive guidance about proceeding to an evaluation of the dog's health 3. Have the RCPC contact the NDDTC to ensure that there is available space for the candidate dog
Fails (Dog did not achieve the minimum mean rating for a given part of the evaluation)	Arrange for the return of the dog to its owners at the owner's expense

Evaluate Health

The health screening must be performed in the following sequence by an accredited and licensed veterinarian. The RCPC or Canine officer regionally approved to procure accompanies the dog to the veterinarian's office for the health screening.

1. General exam
2. Occult heart worm test
3. Blood test
4. Ventro-dorsal x-ray of hips and lateral spinal x-rays
5. Eating habits

Important

At any point in the health screening process, the dog may be eliminated if the results indicate abnormalities.

Use the Health Evaluation Protocol Worksheet to record the results of the health screening. The worksheet is located in **Appendix K** and has been printed so it can be removed, photocopied, and reused.

General Exam

After the general exam is completed, ensure that the veterinarian's findings are within normal limits.

TABLE 2-4-4: Action to Take After a Veterinarian Completes the General Exam

If the findings of the general exam are:	Then:
Outside normal limits	1. Note that the dog is an unacceptable candidate 2. STOP the evaluation process. 3. Arrange for the return of the dog to its owners at the owner's expense
Within normal limits	Continue with the health screening

Heart Worm Test

Request that the veterinarian perform an occult heart worm test.

TABLE 2-4-5: Action to Take After a Veterinarian Completes the Heart Worm Test

If the result of the heart worm test is:	Then:
Positive	1. Note that the dog is an unacceptable candidate. 2. STOP the evaluation process. 3. Arrange for the return of the dog to its owners at their expense.
Negative or within normal limits	Continue with the health screening.

Blood Test

Request that the veterinarian perform pre-surgical blood work or a blood test that includes liver values, kidney values, and a complete blood count. Refer to the Health Evaluation Protocol Worksheet for specific test needs.

TABLE 2-4-6: Action to Take After a Veterinarian Completes the Blood Test

If the results of the blood test are:	Then:
Outside normal limits	1. Note that the dog is an unacceptable candidate. 2. STOP the evaluation process. 3. Arrange for the return of the dog to its owners at their expense.
Within normal limits	Continue with the health screening.

X-rays

Request that the veterinarian perform ventro-dorsal pelvic x-rays and thoracic-lumbar junction spinal x-rays. Hip and spine x-rays must be taken in accordance with positioning guidelines set out by American Veterinary Medicine Association (AVMA). Refer to *Figure 2-4-2*.

Procedures: Procuring Detector Dogs
Evaluate Health

> **NOTICE** The technique established by the Orthopedic Foundation for Animals, Inc. should be used but certification is not required.

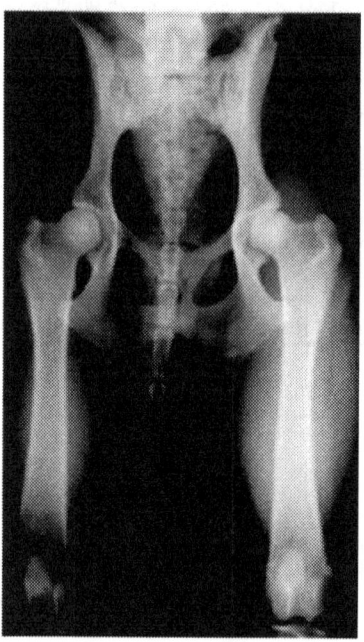

FIGURE 2-4-2: AVMA Positioning Criteria for X-rays

The x-rays must be properly identified and labeled as LEFT and RIGHT.

> **NOTICE** The dog must be anesthetized to perform x-rays.

The NDDTC veterinarian must view the x-rays for final approval. Inform the field veterinarian that the purpose of final approval by the NDDTC veterinarian is to ensure uniformity in the NDDTC standards of approval for hip and spine x-rays.

TABLE 2-4-7: Action to Take After the NDDTC Veterinarian Views the X-Rays

If the x-rays show that:	Then:
Hips are dysplastic or deformed in any way	1. Note that the dog is an unacceptable candidate 2. STOP the evaluation process. 3. Arrange for the return of the dog to its owners at their expense
Within normal limits (OFA rating of fair or above)	Continue with the health screening

Procedures: Procuring Detector Dogs
Complete Final Release Form

Eating Habits

Request that the veterinarian perform an evaluation of the dog's eating habits to determine if there is evidence of kennel stress.

TABLE 2-4-8: Action to Take After a Veterinarian Evaluates the Dog's Eating Habits

If the eating habits of the dog reveal:	Then:
Stress or failure to adjust to the kennel	1. Note that the dog is an unacceptable candidate 2. STOP the evaluation process. 3. Arrange for the return of the dog to its owners at their expense
Normal ability of the dog to adjust to the kennel	Have veterinarian administer the following vaccines: 1. Rabies (one-year vaccine) 2. DHLPP (distemper, hepatitis, leptospirosis, parainfluenza, parvo virus) 3. Corona 4. Bordetella (intra nasal) 5. Fecal examination (internal parasites) **NOTE:** The NDDTC does not require an urinalysis or Lyme disease vaccine; therefore, do not request these

Complete Final Release Form

If the dog successfully passes all aspects of the initial screening, temperament testing, and health evaluation; then, complete a Final Release Form. Refer to *Figure 2-4-3* for a model of the form and to *Appendix K* for a form to copy, which is placed on USDA letterhead.

Procedures: Procuring Detector Dogs
Complete Final Release Form

Subject: Final Release Form

I _____ do hereby relinquish any legal claim
      (Owner)

and/or ownership that I have for _____,
                                       (Name of Dog)

_____ by donation/sale to the U.S. Department
 (Breed of Dog)   (circle one)

of Agriculture for use as a working detector dog. Furthermore, I understand that this dog will be adopted by the public through Federal government procedures upon the retirement of the dog from active duty.

Owner's signature:_____

Date: _____

Location:_____

I _____ request to be given first right of

refusal to the above mentioned dog in the event that it does not pass the

training program. I understand it will be my responsibility to pay

expenses associated with the dog's return.

_____
Owner/Agent

Date: _____

FIGURE 2-4-3: Model of a Final Release Form

Complete Tracking Record and Feedback Worksheet

To monitor the status of the dog's evaluation and training at the NDDTC, complete the top half of a Tracking Record and Feedback Worksheet. The worksheet is located in **Appendix K** and has been printed so it can be removed, photocopied, and reused.

Attach the completed Tracking Record and Feedback Worksheet to the Temperament Evaluation Worksheet.

NDDTC will notify the RCPC with an update to the tracking worksheet for each request submitted.

Complete Airline Flight Tracking Worksheet

Arrange for and ship the dog to the NDDTC using the Airline Flight Tracking Worksheet. The worksheet is located in **Appendix K** and has been printed so it can be removed, photocopied, and reused.

Do not ship dogs on weekends, holidays or outside normal working hours without pre-approval from the NDDTC.

Do not use Acepromazine on any dog being shipped to the NDDTC.

Procedures: Procuring Detector Dogs
Complete Airline Flight Tracking Worksheet

2 Procedures

Public Awareness

National Detector Dog Manual

Contents

Introduction **page-2-5-1**
Demonstrations **page-2-5-2**
List of Outreach Information **page-2-5-2**
Major Media Calls **page-2-5-3**
   Response to Major Media and Congressional Calls **page-2-5-3**
   General Tips for Media Events **page-2-5-4**
Media Relations Training **page-2-5-5**

Introduction

Although public awareness activities are important to USDA's overall mission of protecting American agriculture, they are a secondary function of detector dog teams. Educating the public about the role of the agency is an excellent way to encourage voluntary compliance with regulations that prevent entry of restricted and prohibited fruits, vegetables, and meat through passenger baggage. Since most people find dogs appealing and are impressed by dogs that work, demonstrating the skills of detector dogs has proven to be very effective in promoting the AQI program.

Target audiences include international travelers and the travel industry, school groups, groups connected with the agricultural industry, brokers, and importers.

Important

Canine Officers should contact their RCPCs before participating in public awareness activities that maximize exposure. RCPCs will contact LPA regarding all media events such as newspaper articles, TV, and magazines.

Local managers (port directors and supervisors) should direct the public awareness activities at their work locations and keep RCPCs informed of activities of note.

Procedures: Public Awareness
Demonstrations

Demonstrations

RCPCs are responsible for ordering public awareness information (see the **List of Outreach Information** in this section). Order outreach information and information for demonstrations through RCPCs. RCPCs should allow at least three weeks for processing the request.

Supervisors of Canine Officers are encouraged to take an active role in public awareness activities. Supervisory involvement can enhance appreciation of the role of Canine Officers, and supervisors can assist Canine Officers in presenting a positive and professional perspective of USDA's mission.

Requirements for Selecting Teams

To be selected to represent APHIS' Detector Dog Program at media events, Canine Officers must meet the following requirements:

- Have at least one year with the Detector Dog Program
- Maintain proficiency above 80 percent
- Have permission of their manager
- Maintain a professional appearance
- Be recommended by their RCPC

List of Outreach Information

LPA produces several items and resources to help Canine Officers prepare demonstrations and presentations, depending on the audience. The items and resources are listed below:

- Presentation folders, Don't Pack a Pest
- Coloring books, Miscellaneous Publication No. 1499, Beagle Brigade, Protecting American Agriculture, written in English and Spanish
- PPQ mouse pads
- Beagle Brigade activity sheet
- Bookmarks
- Pamphlets, Miscellaneous Publication No. 1539, USDA's Detector Dogs: Protecting American Agriculture (available online at APHIS' travel website)
- Fact Sheet, Detector Dog Program, February 2003 (available online under News/Publications/Factsheets)

- Beagle Brigade video
- Magnets
- Luggage tags
- Posters
- APHIS' Travel Website http://www.aphis.usda.gov/travel/beagle.html
  - Locations of teams
  - Beagle adoption/donation information

Major Media Calls

All requests from national news media should be forwarded to LPA through the RCPC. LPA's staff coordinates requests with USDA's officials and provides Canine Officers with guidance to ensure updated, accurate, and consistent information. It is advance notification to LPA that is key to the Agency's policy. The speed of electronic communications allows even local stories to be picked up and distributed nationally via news wires and networks.

Major media calls would include those from the following:

- Major daily newspapers, such as USA Today, The Wall Street Journal, Chicago Tribune, New York Times, Los Angeles Times, Albuquerque Journal, Dallas Morning News, the Tennessean, and any Washington metropolitan paper, such as The Journal of Commerce, or The Washington Post
- News magazines, such as Newsweek, Time, and National Geographic
- Network news programs, all network television shows and radio networks, such as NBC, CBS, ABC, Fox, and CNN
- News wire services, such as Associated Press, United Press International, and Reuters

Response to Major Media and Congressional Calls

Follow the steps outlined below when a request is received from the congressional staff or reporters for major media. These steps are found on LPA's Media and Congressional Reference Card.

Never immediately engage in a media discussion.

Important

Procedures: Public Awareness
Major Media Calls

1. Refer all congressional inquiries for PPQ to LPA, Legislative Services at 202-720-2511. Refer inquiries for DHS to Sue Challis at 202-927-1547.
2. Obtain the reporter's name, media affiliation, and phone number.
   A. Determine the topic for the interview or visit.
   B. Find out when the media representative needs the information.
   C. Respond to the request according to APHIS guidelines for responding to the news media (see general tips in this section).
3. Tell the reporter that someone will call them back shortly.
4. Notify your supervisor and call the RCPC, who will call LPA to discuss the request. For media inquiries contact LPA, Public Affairs at 301-734-7799. The regional offices are in California at 916-857-6243; Colorado at 970-494-7410; Florida at 352-332-1893.

General Tips for Media Events

Examples of effective media communication are described in USDA's Office of Communications Guidelines *Media Calls*, January 2002 (refer to *Figure 2-5-1*).

Listed below are general tips on dealing with reporters, community leaders, or members of organizations so you can ensure they provide accurate information to the public about Agency activities and programs.

- Wear a clean, ironed, well-maintained Class A uniform, including a tie and polished shoes
- Bathe the dog prior to the media event
- Make sure the dog's jacket is clean and non-reflective (each team should maintain a jacket for media events)
- Prepare several positive messages; restate them often.
- Remain standing during interview (even if on the phone).
- Know the name and title of your interviewer.
- Be courteous and polite.
- Stay within your field of expertise; never speculate.
- Do not debate.
- Do not justify Agency programs.
- Offer additional information to clarify a story.

- Avoid jargon and technical terms.
- Be aware of time lines. Keep the interview brief—think sound bytes.
- Use precleared information.
- Keep your port informed.
- Never immediately engage in a media discussion.

Media Relations Training

Canine Officers receive training on how to deal with the media during BCOT at NDDTC. Canine Officers receive a copy of the Media Package written by LPA.

USDA United States Department of Agriculture
Office of Communications Guidelines

January 2002

Media Calls

All media calls to USDA must go through public affairs staff at the agencies or the Office of Communications.

Media calls should be given priority and returned ASAP.

The press secretary must approve all requests for on-camera interviews.

Public affairs staff should work directly with the press secretary and communications coordinator on high profile, controversial issues for media response.

Public affairs staff should work with the press secretary and communications coordinator on all public press events.

FIGURE 2-5-1: Media Calls

Procedures: Public Awareness
Media Relations Training

Procedures

Retiring Detector Dogs

Contents

Introduction **page-2-6-1**
Criteria for Retiring a Detector Dog **page-2-6-1**
    Ability of a Detector Dog to Work **page-2-6-1**
    Health Status and History **page-2-6-2**
    When a Canine Officer Leaves the Position **page-2-6-2**
    Aggression **page-2-6-3**
Placing Retired Dogs **page-2-6-3**
    Ranked Options for Placement **page-2-6-3**
    Receipt, Agreement and Waiver of Liability for a Retired Dog **page-2-6-3**
Replacing a Detector Dog **page-2-6-5**

Introduction

The RCPC and the RPM, as representatives of the regional director, are responsible for retiring detector dogs. The RPM must do the following when retiring a detector dog:

- Advise and/or consult with the NDDPM
- Once the decision is made, E-mail the decision to the NDDPM, the NDDTC, and the Port Director

Criteria for Retiring a Detector Dog

The following criteria determine whether a detector dog will continue to work or will retire.

Ability of a Detector Dog to Work

Anyone involved in the management of the canine team may refer questions about the dog's ability to work to the RCPC. The Canine Officer should provide the RCPC with a history of training or work-related problems and measures that have been taken to correct these problems. The NDDTC may be consulted in the assessment of the dog's ability and provided the option of recommending remedial training or alternative duties. Typically, the NDDTC requires training documentation, medical records, and a video tape of the dog working in its port environment for an initial assessment.

Procedures: Retiring Detector Dogs
Criteria for Retiring a Detector Dog

Some patterns of ineffectiveness may include the following:

- Consistently low statistics
- Inability to detect certain odors
- Incompatibility of the team
- Inability to work effectively

Health Status and History

The dog's health must be evaluated by its practicing veterinarian with input from the Canine Officer. If the veterinarian recommends retirement, the recommendation must be in writing before retiring the dog.

A detector dog may be retired because of injury, disease, or age. The following list of examples may be causes for retirement; it is not inclusive.

- Dog reaches nine years of age (Canine Officer will go on the replacement list when the dog reaches 8.5 years of age)
- Hip problems
- Back and neck problems
- Epilepsy
- Arthritis
- Psychological abnormalities
- Mental health problems
- Seizures (zero tolerance)
- Injury
- Skin conditions

When a Canine Officer Leaves the Position

Use **Table 2-6-1** to determine whether a detector dog should be retired when a Canine Officer leaves the position.

Table 2-6-1: Determine Whether to Retire a Detector Dog When a Canine Officer Leaves the Position

If the dog's age is:	And NDDTC:	Then:
1-8 years	Requests the dog	Transfer the dog to NDDTC for reassignment.[1]
	Declines the dog	The RCPC may retire the dog. Go to the options for placing retired dogs beginning on **page-2-6-3**.
9 years	⟶	Retirement is mandatory.

1 NDDTC is responsible for assessing and maintaining all unassigned detector dogs.

Aggression

The dog must be retired if there is one unprovoked, aggressive incident or bite. Do not retire the detector dog until the biting incident has been investigated thoroughly by the RCPC. Investigate all precursory events because when a dog bites someone there was most likely a related event that preceded the bite. Include a medical evaluation in the investigation.

The RCPC will inform the RPM and the NDDPM of the agressive incident or bite. If needed, the NDDPM, in conjunction with the region, is responsible for notifying LPA and PPQ's Deputy Administrator.

Refer to the standard operating procedures for *Dog Aggression* under the *Incidents* section (red tab).

Placing Retired Dogs

Place a detector dog by using the options listed under Ranked Options for Placement. NDDTC may help locate a home for a retiring detector dog.

Ranked Options for Placement

The following options for placement are in order of priority:

1. Canine Officer with most time spent with retiring dog
2. Canine Officer with second most time spent with retiring dog
3. Home placement, such as coworker, friend, or relative
4. Other PPQ personnel in the following order: in the work location, in the area, in the region, and then nationwide. Consult the present Canine Officer in the placement of the dog.
5. If you cannot find a home, refer to the national adoption list. NDDTC may help find a home for a retired dog.

Receipt, Agreement and Waiver of Liability for a Retired Dog

Once a detector dog is placed, then a receipt agreement must be completed. See **Figure 2-6-1 on page-2-6-4** for a sample of an Acknowledgment of Receipt, Agreement and Waiver of Liability. Since 2002, new dogs are identified with USDA microchips by the NDDTC. Send a copy of the surplus dog document to the NDDTC, who will communicate in writing to AKC the transfer to the adopter.

Refer to **Figure 2-6-1** for a sample of an Acknowledgment of Receipt, Agreement and Waiver of Liability.

Procedures: Retiring Detector Dogs
Placing Retired Dogs

When shipping the retired dog, FedEx original copies of its health records to the NDDTC. If the dog is retired within the region, the region will maintain the records.

**ACKNOWLEDGMENT OF RECEIPT, AGREEMENT AND WAIVER OF LIABILITY
(SURPLUS APHIS DOG)**

The Animal and Plant Health Inspection Service (APHIS) has in its possession a surplus dog named _____. Ownership of this dog is hereby transferred to _____, who hereby agrees to the following:

1. I acknowledge receipt of the dog named herein and acknowledge that the United States of America, acting through the U.S. Department of Agriculture, relinquishes all rights, title, and interest in the dog and all responsibility for its condition and actions. I accept and assume full ownership of said dog.

2. In accepting full ownership of said dog, I assume complete responsibility for its condition and actions. I agree that the United States of America, U.S. Department of Agriculture, has no liability for damages to any property or any personal injury, including death, to any person arising from or incident to the donation of the dog or its subsequent use or disposition.

3. I state that I have not exchanged money or anything of value except this agreement and waiver of liability for said dog.

4. I agree not to sell said dog and agree not to place the dog in trade or commerce, in any manner, at any time.

5. Further, accept this dog fully aware that the United States of America, U.S. Department of Agriculture, makes no warranty or guarantee of its physical condition, temperament or future behavior.

6. I agree to purchase or otherwise acquire in my name any and all required licenses within 72 hours of receiving the dog.

Signed this _____ day of _____, 19____.

_____
Signature of Donee

_____
Printed Name of Donee

_____
Address of Donee

FIGURE 2-6-1: Sample of Acknowledgment of Receipt, Agreement and Waiver of Liability

Procedures: Retiring Detector Dogs
Replacing a Detector Dog

Replacing a Detector Dog

The priority for assigning replacement dogs will be as follows:

1. New handlers within 6 months of graduating from BCOT or replacement training
2. Needs of the port (Does the port already have a dog team?)
3. Degree of pest risk at the port
4. Compatibility of the available dog to the handler
5. Compatibility of the dog to the port in question (Will the dog work better at a slow or fast paced port? What is the type of work to be done at the port?)

Important

All Canine Officers will return to NDDTC for replacement training. The length of training will be determined by the NDDTC and the RCPC.

Procedures: Retiring Detector Dogs
Replacing a Detector Dog

2 Procedures

Utilizing Detector Dogs

National Detector Dog Manual

Contents

Implementation Process **page-2-7-2**
Establishing Work Location Operating Procedures **page-2-7-3**
  Flight Selection **page-2-7-3**
  Tours of Duty **page-2-7-3**
  Proficiency Training **page-2-7-4**
  Housing in a Secondary Residence **page-2-7-4**
  Utilizing Down Time **page-2-7-4**
  Exercise and Biological Breaks **page-2-7-5**
  Home Stay **page-2-7-5**
  Approach to Screening **page-2-7-5**
TDY Assignments, Developmental Assignments, and Extended Leave Policy **page-2-7-6**
Courtesy of the Port **page-2-7-7**
Guidelines Agreed to Between APHIS and Customs **page-2-7-7**

Procedures: Utilizing Detector Dogs
Contents

Implementation Process

Refer to **Table 2-7-1** for a quick reference to the main steps in implementing a new detector dog team and determining who is responsible for taking the steps.

TABLE 2-7-1: Implementation Process for a New Detector Dog Team

Steps of the implementation process:	Person(s) responsible for implementing the step:
1. Conduct feasibility study and on-site assessment	Port Director and the RCPC
2. Determine staffing need based on results of feasibility study	Regional Director with input from the RCPC
3. Initiate request to OPM to announce vacancy	Port Director
4. Forward a list of eligible applicants for interview	APHIS Business Services
5. Interview applicants	Port Director with input from the RCPC
6. Select a detector dog officer	Port Director with input from the RCPC
7. Request training	Port Director through the RCPC 2 months in advance
8. Test and procure detector dogs	National Canine Instructor, RCPC, or designee
9. Orient Canine Officer at work location. Included in this orientation is Canine Officers completing port and PPQ 436 officer requirements before attending training.	Port Director, supervisor
10. Conduct protocol training of detector dogs at NDDTC (5 weeks)	National Canine Instructor
11. Conduct BCOT at NDDTC (10 weeks)	National Canine Instructor
12. Conduct installation assessment at work location (2–6 weeks after BCOT)	RCPC, coordinated with Port Director
13. Conduct follow up contact (6 months after BCOT)	RCPC (consulting with National Canine Instructor)
14. Conduct validation testing, annually or as necessary	RCPC
15. Prepare statistical reports monthly or more frequently as required by the program	Canine Officer, Port Director, RCPC

Procedures: Utilizing Detector Dogs
Establishing Work Location Operating Procedures

Establishing Work Location Operating Procedures

All parties involved including Port Directors, supervisors, Canine Officers, and RCPCs should establish the operating procedures for a detector dog team. Procedures are designed for individual work location situations. The following topics should be considered when establishing the work location operating procedures.

Flight Selection

In order to select flights that will best utilize the detector dog and enhance pest exclusions, the RCPC, the Canine Officer, and the supervisor must work together. Work with the RCPC to establish a work schedule for detector dog teams that will take advantage of international traffic (flights, mail, cargo) that best uses detector dogs, based on the results of feasibility studies, port records such as PPQ 212s, WADS, AQI monitoring data, pest risk, and country risk (high, medium, or low), and other port activities. With Canine Officers determine targeted flights within this time frame.

Selected flights should be continuously reviewed by the Canine Officer and evaluated by the RCPC. Environmental impacts, including growing seasons, traveling cycles, and origin of the carrier, can change the success rate of a detector dog team on a particular flight. The Canine Officer is responsible for continually evaluating risk by working detector dogs on a variety of flights.

The Canine Officer provides assistance in defining the flights on which the detector dog team would be more effective in finding contraband. Additional assistance can be provided by the RCPC.

Tours of Duty

Schedule a detector dog team so that it is present when most needed. All scheduled tours of duty should be considered in determining the most effective use of assigned detector dog teams. Detector dog team tours do not have to coincide with existing tours of duty. To maintain its health and welfare, the detector dog must have a scheduled day off each week.

The Canine Officer should be allotted sufficient time to return the detector dog to the primary residence (boarding kennel) before any 436 officer overtime is undertaken as a PPQ 436 officer. If this is not possible, certain precautions must be taken to ensure that the health and safety of the dog are not compromised, the AWA standards are met, and the maximum amount of time a dog can be housed in a secondary residence is not exceeded. (See *Housing in a Secondary Residence*).

Proficiency Training

Proficiency training is an extension of initial training. During proficiency training, the canine officer uses many of the same procedures used during initial training, yet their objectives may differ. The difference is usually in the complexity of the problem in correcting deficiencies versus the dog learning something new. You cannot rule out the latter.

Proficiency training must be conducted on a continuous basis. The Canine Officer establishes a schedule (along with local managers) for conducting training based on the needs of his or her particular dog. Some dogs will require more training than others.

The best gauge of the amount of training to conduct is the dog's efficiency. If the dog has no deficiencies and is making the normal amount of actual finds, the amount of proficiency training needed can be limited to increasing the dog's sensitivity. If the reverse is true, the canine officer should consider additional training time. Four hours per week is sufficient for the average dog, though it should be kept in mind that additional time is needed to set up the required training problem.

Housing in a Secondary Residence

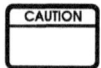 A detector dog must not be housed longer than 12 consecutive hours in its secondary residence (crate or wire kennel), unless it is being shipped in a portable kennel. Refer to **Appendix E** for guidelines on how to ship detector dogs. Refer to the **Glossary** for the definition of a secondary residence.

Utilizing Down Time

Tours of duty of Canine Officers should be scheduled so there is minimal down time. During this down time, other canine-related duties should be performed, such as preparing required monthly and training reports, grooming, exercising, training, and other responsibilities directly related to detector dog activities.

If no detector dog duties are being performed, the Canine Officer should then perform PPQ 436 officer duties, such as secondary inspection, pest identification, and record keeping.

 When Canine Officers are performing other duties, the detector dog must be confined in its secondary residence, not tied to a stationary object. Canine Officers cannot allow detector dogs to do any of the following:

- Roam around the office or common areas
- Mingle with nondetector dogs
- Move about an unsecured area off a leash

Procedures: Utilizing Detector Dogs
Establishing Work Location Operating Procedures

Exercise and Biological Breaks

Following are health guidelines to consider while detector dogs are working and being transported. These guidelines comply with the requirements of the Animal Welfare Act (AWA).

Exercise

For the health and welfare of the detector dog, it should be exercised.

Exercising a detector dog off leash with other dogs can increase the risk of disease and injury. Because of the physical demands of canine working activities, adverse health conditions may be created or aggravated by play exercises with other ag working dogs. Back injuries are of great concern. To prevent future back injuries, all dogs over the age of 5 or dogs that have suffered any back related injury must have medical clearance for play activities with other working dogs.

Biological Breaks

While a detector dog is housed in a secondary residence (crate or wire kennel), allow a biological break at least every 2 hours. Take the detector dog to an area where it can urinate and/or defecate. **Plan for at least a 15 minute biological break.** This time line is a guide. Some detector dogs may require more frequent biological breaks, based on their individual functions.

Home Stay

Canine Officers cannot take detector dogs home except in extraordinary circumstances (i.e., medical situation, dog on medication, recovery period after surgery). Approval to take a dog home must be granted by the RPM upon consultation with the RCPC and the Port Director.

Approach to Screening

Detector dogs should only be used by their assigned handlers, except in extraordinary cases with prior approval of the RCPC. Detector dogs can be used only in areas in which they have been trained, and large breed active response dogs cannot be used for passenger clearance in FIS areas.

Detector dog teams should be used to screen passengers in an area that allows maximum exposure to baggage (i.e., baggage carousel, exit points). At post offices, detector dogs should work where they have access to most packages being released from the FIS area. The screening approach should be left up to the discretion of the Canine Officer, in consultation with local port management.

Procedures: Utilizing Detector Dogs
TDY Assignments, Developmental Assignments, and Extended Leave Policy

The following are general steps Canine Officers take once a detector dog responds to passenger baggage:

1. Note the response on the passenger's declaration form. This step ensures that suspect baggage is directed to PPQ personnel and is identified as a detector dog response.

2. **Visually inspect hand-carried baggage.** It is necessary to verify the accuracy of responses and it is an integral part of the detector dog inspection process. This step may include safeguarding contraband, and it frequently leads to improved detector dog proficiency and passenger processing.

Detector dog teams with over one year of experience in the field must maintain a proficiency rating of at least 80%.

Important

TDY Assignments, Developmental Assignments, and Extended Leave Policy

Canine Officers are allowed to be on rapid response teams and to participate in TDY assignments. They should not be denied a TDY assignment solely because they are Canine Officers.

The only restrictions about Canine Officers taking TDY assignments are that there will be none scheduled within:

- The first year after graduating from BCOT
- Within 6 months after attending replacement training

Canine Officers should contact their Port Director and RCPC when assigned to or requesting a TDY assignment, a developmental assignment, or an extended leave of absence.

The Port Director works with the RCPC to ensure the detector dog's proficiency remains consistent while the Canine Officer is away. Use **Table 2-7-2** to determine the action to take to ensure detector dog proficiency.

Procedures: Utilizing Detector Dogs
Courtesy of the Port

TABLE 2-7-2: Action to Take to Ensure Consistent Proficiency of a Detector Dog When a Canine Officer Is Away for Extended Periods

When a Canine Officer will be away for:	Then do the following:
30 days to 6 months	1. Temporarily reassign the detector dog. The RCPC is responsible for reassigning the dog within the region. 2. If 1. is impossible, the RCPC contacts the RPM, who in turn contacts the NDDPM for help reassigning the dog to another region or at NDDTC. 3. The RCPC provides on-site support to facilitate a smooth transition back into a productive detector dog team when the Canine Officer returns. 4. As a result of observations by the RCPC and Port Director, they can request technical support from NDDTC. 5. Regional management may, at their discretion, discuss alternatives with the NDDTC staff.
More than 6 months	The RCPC notifies the RPM and the NDDPM to facilitate resource scheduling, including reassignment of the dog.

Courtesy of the Port

Detector dog teams are more likely to encounter diplomats when clearing a flight as opposed to regular PPQ 436 officers without the assistance of a detector dog.

As a Canine Officer, when diplomats are encountered, refer to the Airport and Maritime Operations Manual, Airport, Clearing Passengers/Crew for guidelines about courtesy of the port. If there is a positive response on a diplomatic bag, the Canine Officer should request the diplomat's permission to inspect the bag. Also, be aware of additional port policies for diplomats.

Guidelines Agreed to Between APHIS and Customs

The following guidelines are taken from an APHIS Detector Dog Agreement established with the U.S. Customs Service. Refer to these guidelines, along with port policy, when developing local guidelines for cooperatively working with detector dogs at a work location.

1. When a detector dog responds to handbaggage, Canine Officers will examine it on the spot. If something is found that requires a referral to secondary inspection, place an "A" on the passenger's declaration card. Canine Officers may remove small amounts of prohibited items at the baggage carousel or elsewhere on the floor.

Procedures: Utilizing Detector Dogs
Guidelines Agreed to Between APHIS and Customs

2. When a detector dog responds to pit baggage, the Canine Officer will place an "A" on the passenger's declaration card and will refer the passenger to U.S. agricultural secondary inspection.

3. Canine Officers will direct passengers to secondary inspection after a positive alert.

4. U.S. Customs' personnel will be given instructions by local APHIS managers on APHIS procedures.

Health Care

Anatomy

Contents

Introduction **page-3-1-1**
External Anatomy of a Beagle **page-3-1-2**
Internal Anatomy of a Beagle **page-3-1-3**

Introduction

Use this section of the manual to identify and describe the external and internal anatomy of a dog. Knowing the terms used to describe a dog's anatomy enables Canine Officers to more efficiently report problems to veterinarians. ***FIGURE 3-1-1*** shows the external body parts. ***FIGURE 3-1-2*** shows the skeletal structure, while ***FIGURE 3-1-3*** shows the male internal organs and ***FIGURE 3-1-4*** shows the female internal organs.

Health Care: Anatomy
External Anatomy of a Beagle

External Anatomy of a Beagle

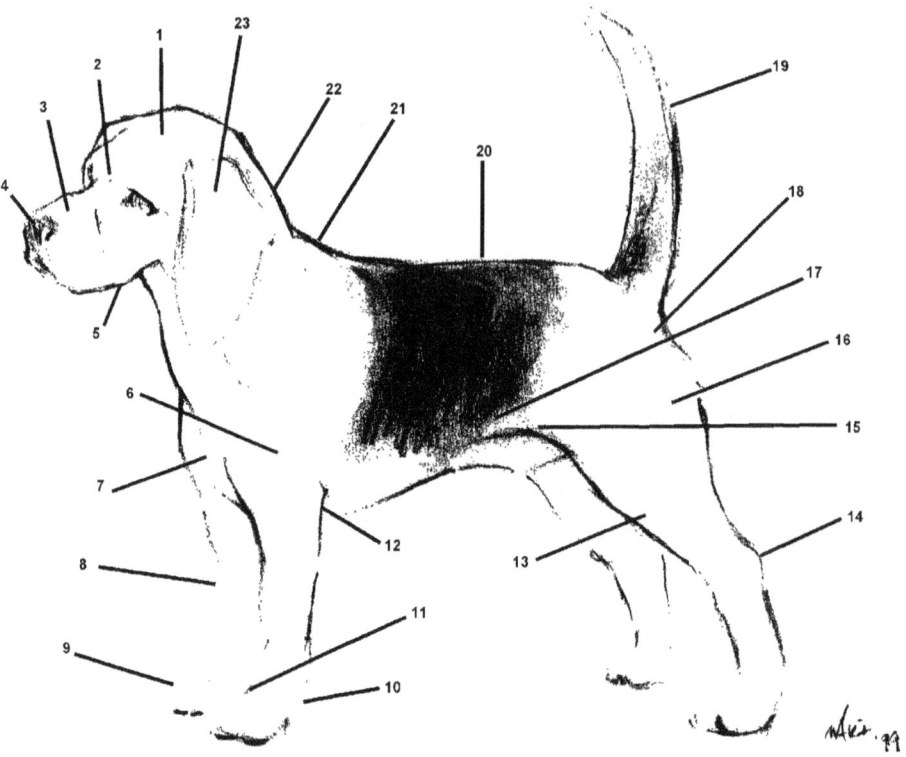

1. Forehead	9. Foot	17. Abdomen
2. Stop	10. Pastern	18. Rump
3. Muzzle	11. Dewclaw	19. Tail
4. Nose	12. Elbow	20. Back
5. Lower jaw	13. Stifle	21. Withers
6. Shoulders	14. Hock	22. Neck
7. Chest	15. Flank	23. Ear
8. Foreleg	16. Thigh	

FIGURE 3-1-1: External Anatomy of a Beagle

Health Care: Anatomy
Internal Anatomy of a Beagle

Internal Anatomy of a Beagle

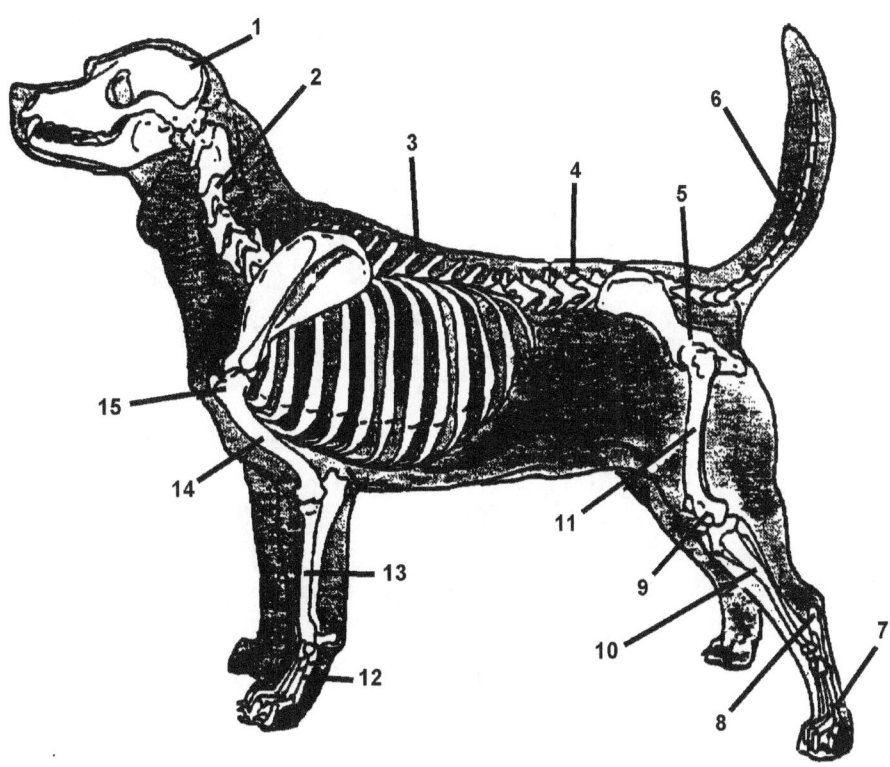

1. Occiput	6. Tail vertebrae	11. Upper thigh (femur)
2. Cervical vertebrae	7. Back foot or metatarsus	12. Pastern or metacarpus
3. Withers	8. Tarsus or hock	13. Forearm (radius and ulna)
4. Vertebrae column	9. Stifle or knee	14. Upper-arm (humerus)
5. Hip joint	10. Lower thigh (tibia and fibula)	15. Shoulder joint

FIGURE 3-1-2: Internal Anatomy of a Beagle—Skeletal Structure

Health Care: Anatomy
Internal Anatomy of a Beagle

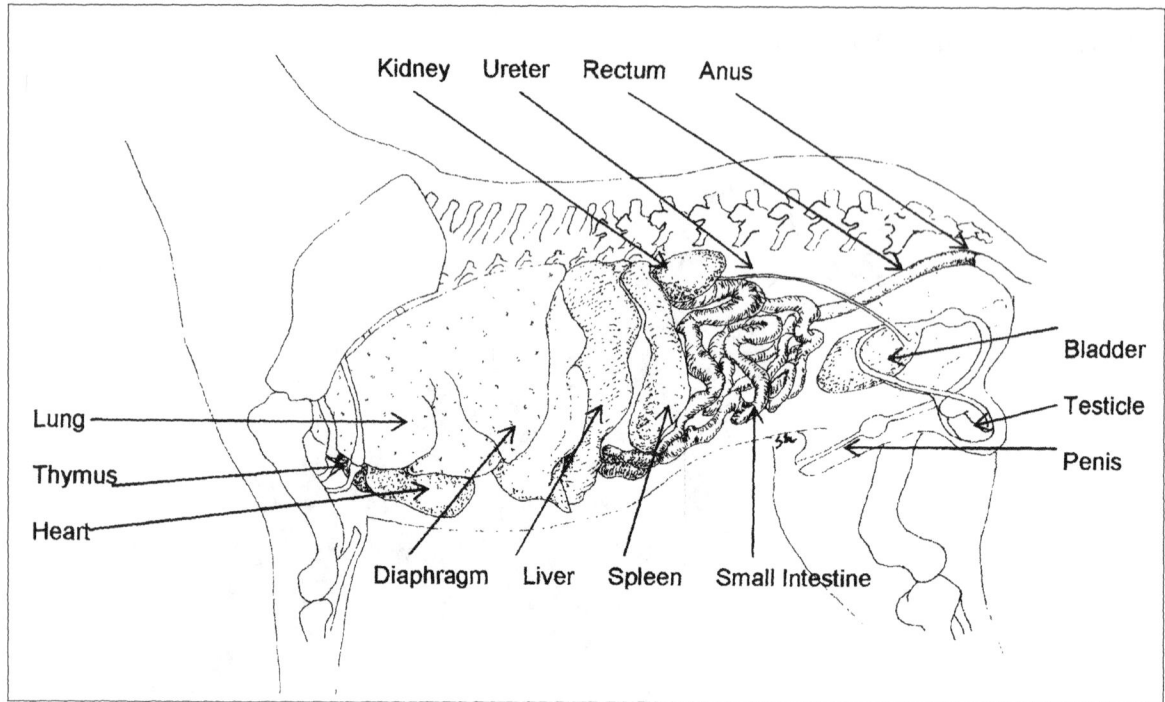

FIGURE 3-1-3: Internal Anatomy of a Beagle—Internal Organs (male)

Health Care: Anatomy
Internal Anatomy of a Beagle

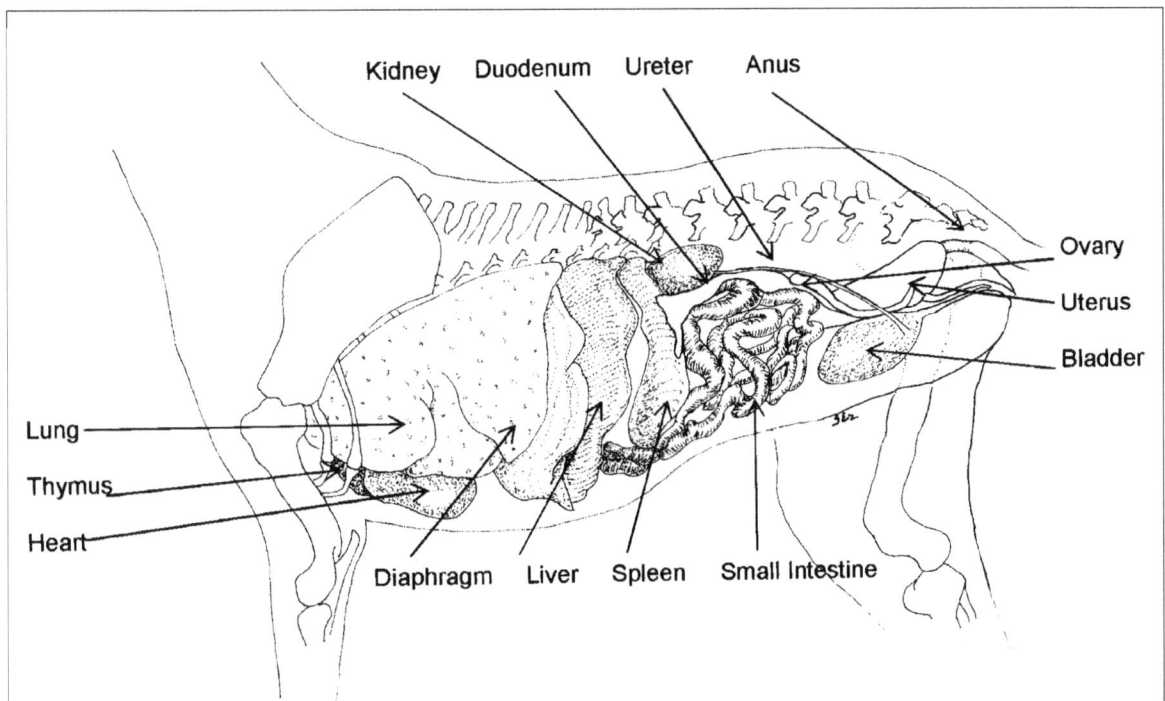

FIGURE 3-1-4: Internal Anatomy of a Beagle—Internal Organs (female)

Health Care: Anatomy
Internal Anatomy of a Beagle

Health Care

Diseases and Parasites

Contents

Introduction   page 3-2-1
Infectious Diseases   page 3-2-2
   Disease Prevention and Vaccination   page 3-2-2
   Distemper   page 3-2-2
   Hepatitis   page 3-2-3
   Leptospirosis   page 3-2-3
   Lyme Disease   page 3-2-3
   Rabies   page 3-2-4
   Parvo Virus   page 3-2-4
   Rocky Mountain Spotted Fever   page 3-2-4
   Corona Virus   page 3-2-5
   Canine Infectious Tracheo-bronchitis (CITB)   page 3-2-5
   Summary   page 3-2-6
Noninfectious Diseases and Severe Illnesses   page 3-2-8
Parasites   page 3-2-8
   External Parasites   page 3-2-8
     Fleas   page 3-2-8
     Lice   page 3-2-9
     Mites   page 3-2-9
     Ticks   page 3-2-9
   Internal Parasites   page 3-2-11
     Heartworms   page 3-2-11
     Hookworms   page 3-2-12
     Roundworms   page 3-2-13
     Tapeworms   page 3-2-14
     Whipworms   page 3-2-15
     Summary   page 3-2-15

Introduction

Taking proper care of a detector dog is extremely important to its overall health and will greatly affect its performance. Canine Officers are responsible for the well being of their detector dogs. Areas to be aware of are the following:

- Safety hazards
- Disturbing influences that may interfere with the dog's rest and relaxation
- Adequate kennel construction
- Climatic conditions
- Feeding and watering schedules

Health Care: Diseases and Parasites
Infectious Diseases

Veterinarians are the best source for information about the health and care of dogs, feeding and watering schedules, and first aid procedures. Establish a good working relationship with the veterinarian responsible for the care of the detector dogs.

Infectious Diseases

Infectious diseases are caused by microorganisms that can be transmitted without actual contact. Diseases that are transmitted from one animal to another are called contagious. Diseases that are transmitted from an animal to a human are called zoonotic.

Disease Prevention and Vaccination

Because of the importance and the nature of their demanding work, detector dogs must stay healthy. Veterinarians are experts in treating, preventing, and controlling diseases that could infect detector dogs. Canine officers help prevent diseases by becoming familiar with their dog's normal body functions, such as appetite and stool, and by knowing when to report potential medical problems. Therefore, canine officers must be familiar with the symptoms of diseases and parasites in order to detect potential problems.

The important diseases that can infect detector dogs are listed and then described below. These diseases are prevented by annual vaccinations.

- Distemper
- Hepatitis
- Leptospirosis
- Lyme disease bacteria (in areas where prevalent)
- Rabies (1 or 3 year vaccine)
- Parvo virus
- Rocky Mountain Spotted Fever
- Corona virus (Canine Conaviral Enteritis)
- Canine Infectious Tracheobronchitis (CITB)

Distemper

Distemper is a widespread, highly contagious, and often fatal disease that occurs primarily in young dogs. The airborne virus is easily transmitted from dog to dog, but does not affect humans.

Hepatitis

Hepatitis is a widespread, viral disease mostly found in young dogs, but can also infect older dogs that have not been immunized. It does not infect humans. Infectious hepatitis is spread through urine—primarily when feeding and drinking utensils are contaminated with urine. The mortality rate is not as high as is from distemper, but recovery takes a long time.

Leptospirosis

Leptospirosis is commonly known as lepto. It is caused by a spirochete (bacterium of the order Spirochaetales) and is fairly common. Other animals can be infected by this disease, and it can be transmitted to humans. It is spread through the urine, usually of dogs and rats. Rodent control is important in preventing the spread of lepto since rats are a common carrier.

 Rodenticides are highly toxic to dogs!

It is essential that dogs do not consume contaminated food or water. Where lepto is known to exist or suspected to exist, dogs should not be allowed to enter or drink surface water that may be contaminated by urine or dead animals. The possibility of human infection reinforces the need for personal cleanliness. Canine officers must protect themselves from urine or contamination when caring for sick dogs.

Lyme Disease

Lyme disease is an infection caused by a bacteria. The disease is spread by the bite of ticks that are infected with the bacteria. Typically, the larvae and nymphs become infected with Lyme disease bacteria when they feed on infected small animals then, infected nymphs and adult ticks bit and transmit Lyme disease bacteria to other animals and humans.

Lyme disease is difficult to diagnose because its symptoms are similar to other diseases, such as fever, loss of appetite, swelling of the legs, joint and muscle pains, and staggering gait. Even though Lyme disease is treatable, some dogs that have become infected with the disease have developed arthritis.

Ticks search for host animals from the tips of grasses and shrubs and transfer to an animal (or person) that brushes up against the vegetation. They frequent wooded, brushy, and grassy places. The risk of exposure to ticks is greatest in the woods and landscaped areas of properties. Therefore, it is particularly important to take preventive action in those areas where Lyme disease occurs. Lyme disease occurs along the east coast from Maine to Florida; in the north central States, especially Wisconsin and Minnesota; in the south from Alabama to Texas; and along the west coast.

Health Care: Diseases and Parasites
Infectious Diseases

When selecting a kennel, ensure that outdoor runs and fenced-in areas are well maintained with leaves removed, and brush and tall grass trimmed away from the buildings and edges of the runs. Ensure that the kennel has a treatment plan to prevent tick infestations.

Canine officers should take extra care when checking their dog's skin for the presence of ticks and fleas during the spring through early fall, especially if they are located in areas where Lyme disease is known to occur. The male tick is a small, flat insect about the size of a match head. The female tick, which is the blood feeder, may swell up to the size of a pea. Both are attached to the dog only by their mouth parts. See additional information about how to remove ticks under *Ticks* in this section.

Rabies

Rabies is an acute, infectious, often fatal viral disease of most warm-blooded animals. It attacks the central nervous system. It is transmitted by the bite of an affected animal, or by contact with the saliva of an affected animal with broken skin. The animals most frequently affected are skunks, raccoons, bats, foxes, dogs, cattle, and cats.

Symptoms may include a sudden change in temperament or attitude, excitement, difficulty in swallowing water or food, blank expression, slack jaw, excessive drooling, paralysis, coma, and death. Wild animals with rabies often lose their natural fear and attack rather than retreat.

Canine officers must prevent contact between detector dogs and wild or stray animals. Report to the veterinarian any contact resulting in a bite or scratch. Use extreme caution while capturing an animal to prevent bites to humans. Medical treatment should be given as soon as possible if a canine officer is bitten by an animal.

Parvo Virus

Parvo virus is a highly contagious disease that causes diarrhea and vomiting and often can be fatal. The highest mortality rate is in dogs less than 12 weeks old. In adult dogs, the symptoms are usually less severe, resulting in fewer deaths.

Other symptoms of the disease include passing or vomiting blood followed by rapid dehydration. Sometimes dogs infected with Parvo virus have jaundice.

Rocky Mountain Spotted Fever

Rocky Mountain spotted fever is a disease transmitted by several species of ticks. The disease occurs in the east from New York to Florida, and in the south from Alabama to Texas and is more frequently seen from April through September.

Infected adult ticks transmit the disease to dogs during biting and feeding. The symptoms in dogs include listlessness, conjunctivitis, depression, high fever, loss of appetite, cough, difficult breathing, swelling of the legs, joint and muscle pains, vomiting and diarrhea, staggering gait, altered mental state, and seizures. These symptoms are similar to distemper, which may be the first diagnosis.

Canine officers should take extra care when checking their dog's skin for the presence of ticks and fleas during the spring through early fall, especially if they are located in areas where Rocky Mountain spotted fever is known to occur. See additional information about how to remove ticks under *Ticks* in this section.

Corona Virus

Corona virus is a highly contagious disease that causes diarrhea and vomiting. It is the second leading cause of viral diarrhea next to Parvo virus. Corona virus weakens the dog by causing severe diarrhea, vomiting, excessive thirst, weight loss, listlessness, and loss of appetite. It affects dogs of all ages, but it severely affects puppies. It is also possible for dogs to be affected by both Parvo virus and Corona virus at the same time.

Canine Infectious Tracheo-bronchitis (CITB)

CITB is commonly known as kennel cough and is usually self-limiting and is rarely fatal. It is usually a mix of viral and bacterial agents. The most common viruses involved are parainfluenza and canine adenovirus. Other factors, including mycoplasmas and canine distemper, can cause severe and potentially fatal complications, such as pneumonia.

CITB is an airborne infection with an incubation period of 5–10 days. It appears in two main forms. The milder form lasts 1–3 weeks and occurs most often in dogs that have been vaccinated against distemper and hepatitis. Symptoms include a dry, hacking cough that might be followed by retching and vomiting. In some cases, pneumonia might follow the mild disease. The severe form is more common in dogs with an uncertain vaccination history. It starts with a dry, mucoid, and sometimes painful cough that can progress to severe bronchopneumonia. In some cases, the severe form could be fatal.

CITB is a highly contagious disease striking even the cleanest, best operated kennel facilities. In order to prevent infection, local immunity must be created in the respiratory tract. An intranasal vaccination (like nose drops) must be administered by a veterinarian to provide safe, effective protection from the disease.

Health Care: Diseases and Parasites
Infectious Diseases

Summary

Refer to **Table 3-2-1** for a summary of the important diseases that could affect detector dogs. The table summarizes the symptoms to look for when observing dogs and kennel facilities. Canine officers should notify the veterinarian when a detector dog shows one or more of the symptoms of diseases.

Other diseases for which vaccines do not exist can affect dogs, such as, upper respiratory infections, pneumonia, and gastroenteritis. Affected dogs may show symptoms including high temperature, loss of appetite, loss of energy, vomiting, diarrhea, and coughing. Any of these symptoms should be reported to a veterinarian.

Table 3-2-1: Summary of Important Infectious Diseases

Name of disease:	Symptoms:	Caused by:	Transmitted by:	Prevention method:
Distemper	◆ Fever ◆ Loss of appetite ◆ Depression ◆ Discharge from eyes and nose	Virus	Airborne or by direct contact	Distemper vaccination annually
Hepatitis	◆ Fever ◆ Loss of appetite ◆ Depression ◆ Discharge from eyes and nose	Virus	Urine	◆ Hepatitis vaccination annually ◆ Practice good sanitation in kennel
Leptospirosis	◆ Fever ◆ Loss of appetite ◆ Diarrhea ◆ Vomiting	Bacteria (Spirochete)	Urine	◆ Leptospirosis vaccination annually ◆ Practice good sanitation in kennel
Lyme disease	◆ Fever ◆ Loss of appetite ◆ Swelling of legs ◆ Joint and muscle pain ◆ Staggering gait	Bacteria (Borrelia burgdor Feri)	Bite and feeding of infected tick	◆ Practice good sanitation in kennel ◆ Well-kept grounds of kennel and runs ◆ Veterinarian prescribed vaccination in prevalent areas ◆ Daily health checks of dog's skin, especially in the spring through fall
Parvo virus	◆ Diarrhea ◆ Vomiting ◆ Passing of blood ◆ Dehydration (continued on next page)	Virus	Airborne or by direct contact	◆ Minimal annual vaccination ◆ Parvo virus vaccination annually

Health Care: Diseases and Parasites
Infectious Diseases

Table 3-2-1: Summary of Important Infectious Diseases (continued)

Name of disease:	Symptoms:	Caused by:	Transmitted by:	Prevention method:
Rocky Mountain spotted fever	◆ Listlessness ◆ Conjunctivitis ◆ Depression ◆ High fever ◆ Loss of appetite ◆ Cough ◆ Difficult breathing ◆ Swelling of the legs ◆ Joint and muscle pain ◆ Vomiting and diarrhea ◆ Staggering gait ◆ Altered mental state ◆ Seizures	Rickettsial disease	Bite and feeding of infected tick	◆ Practice good sanitation in kennel ◆ Well-kept grounds of kennel and runs ◆ Daily health check of dog's skin, especially in the spring through fall
Corona virus	◆ Diarrhea ◆ Vomiting ◆ Excessive thirst ◆ Loss of appetite ◆ Listlessness ◆ Weight loss	Virus	Airborne or by direct contact	◆ Minimal annual vaccination ◆ Corona virus vaccination annually
Rabies	◆ Change in temperament ◆ Difficulty swallowing ◆ Blank expression ◆ Slack jaw ◆ Excessive drooling ◆ Seeks solitude ◆ Depression ◆ Paralysis ◆ Coma and death in 7-10 days	Virus	Saliva from infected animal	Rabies vaccination (1 or 3 year vaccine)
CITB (kennel cough)	◆ Fever ◆ Runny nose ◆ Red or watery eyes ◆ Dry, hacking cough ◆ Retching and vomiting	Virus	Airborne or by direct contact	◆ Minimal annual vaccination ◆ Intranasal vaccination annually

Noninfectious Diseases and Severe Illnesses

Many illnesses affecting dogs are not caused by viruses, bacteria, or other infectious diseases. Noninfectious diseases and severe illnesses include: overheating, arthritis, bloating, chronic kidney disease, tick paralysis (see **Ticks** under *External Parasites*), epilepsy, and allergies.

Since symptoms of a noninfectious disease or severe illness may resemble those of an infectious disease, canine officers should note any abnormality, such as a gradual loss of weight, excessive urination, and obscure lameness. They should report their observation to a veterinarian.

Parasites

Parasites are organisms that infest a host animal for the purpose of feeding from the host's body. Most parasites are harmful to a dog's health, and some parasites can spread diseases to other dogs or humans. Dogs may serve as hosts to a large number of parasites; therefore, controlling parasitic infestations is very important.

External Parasites

External parasites live in or on the skin of the dog. They cause damage by sucking blood or actually eating the tissue. The dog responds by biting and scratching the irritated areas, which may lead to severe skin infections and reduce the dogs working capabilities. The most common external parasites are listed and then described below.

- Fleas
- Lice
- Mites
- Ticks

Fleas

Fleas are small, wingless, bloodsucking parasites of warm-blooded animals. They have legs adapted for jumping. Fleas torment a dog, irritate their skin, and spread disease. They crawl or jump very rapidly through the dog's coat.

Like ticks, fleas are difficult to control since they do not spend all of their time on the dog's body, but live in cracks in the kennels. Fleas may also transmit tapeworms from dog to dog. Controlling fleas requires repeated, individual treatment and continuous kennel sanitation.

Health Care: Diseases and Parasites
Parasites

Lice

Lice (plural for louse) are numerous small, flat-bodied, wingless biting or sucking insects, many of which are external parasites of dogs and humans. Biting lice live off the dog's tissues, while the sucking lice live off the dog's blood. Both produce great irritation for a dog. Biting lice crawl over the skin and through the hair. Sucking lice are usually immobile, and stand perpendicular to the skin. The eggs of lice, called nits, are small, white or gray, crescent-shaped objects fastened to the hairs. Lice, unlike fleas and ticks, can only live a short time when they are not on the dog's body. Controlling lice requires treatment only of infested animals.

Mites

Mites of several types irritate the dog's ear canal (ear mite) or produce mange (mange mite). Mites spend their entire life on the dog. Controlling mites depends primarily on treating the infested dog.

The ear mite lives in the ear canal and causes a severe irritation. The mites are small but are visible to the naked eye as tiny, white crawling specks. Affected dogs scratch at the ear and cock their heads to one side, or shake their heads. Examine the ear canals for a large amount of dark-colored waxy discharge.

Mange mites live in the dog's skin. The sarcoptic mange mite can be transmitted to humans. Mange mites are too small to be seen by the naked eye, but a skin scraping of the infested area will reveal them under a microscope. Canine officers should watch for unusual hair loss as a sign of mange mites.

Ticks

Ticks are small, bloodsucking parasites. They are common in many parts of the world. Ticks feed on blood by inserting their mouth parts into the skin of a host animal. When they are present in large numbers, they can cause a serious loss of blood. Ticks spread diseases by feeding on blood or tissue fluid from a diseased animal and then moving to another animal. Ticks search for host animals from the tips of grasses and shrubs and transfer to an animal (or person) that brushes up against the vegetation. Ticks only crawl; they do not jump or fly.

A noninfectious disease transmitted by ticks is called tick paralysis. Ticks are capable of secreting a toxin that causes paralysis in dogs. Not all infected animals become paralyzed. The adult ticks of some species produce a salivary toxin that enters the blood stream of a host animal and interferes with its nervous system. The onset of symptoms is gradual, with paralysis affecting the pelvis area first, resulting in a staggering gait. Other early symptoms include an altered or impaired voice and cough. Within 24-72 hours, a dog lies down, its reflexes diminish, its jaw muscle weakens, and facial paralysis is noticeable. Death may occur within several days from respiratory paralysis. Recovery is usually good and occurs within 1–3 days after removing the tick and/or treating the dog.

Health Care: Diseases and Parasites
Parasites

Two other important diseases transmitted by ticks are Lyme disease bacteria and Rocky Mountain spotted fever. See additional information about these two diseases in this section.

Canine officers should take extra care when checking their dog's skin for the presence of ticks during the spring through early fall, especially if they are located in areas where Lyme disease bacteria and Rocky Mountain spotted fever are known to occur. Ticks are usually found on the ears, neck, head, and between the toes of a dog. The male tick is a small, flat insect about the size of a match head. The female tick, which is the blood feeder, may swell up to the size of a pea. Both are attached to the dog only by their mouth parts.

Canine officers must be very careful when removing ticks, since they may carry diseases transmittable to humans. Also, if all of the tick is not removed, the skin may become inflamed. For ticks attached deep within the ear canal, have them removed by a veterinarian to avoid injury to the dog's ear. To remove a tick, place your index finger and thumb nails (or tweezers) around the body of the tick as close to the dog's skin as possible. Slowly withdraw the tick's head from the skin. Flush the tick down the nearest drain, or immerse it in alcohol.

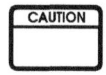 Do not crush or squeeze a tick with your bare fingers.

 After handling ticks, wash your hands with soap and water because they can carry diseases that are transmittable to humans.

Ticks may be found in cracks in the floors and sides of the kennel and in the grass and bushes in the training areas. They may live away from the dog's body as long as a year without having to return for a blood meal. To control ticks, the kennels, training areas, and working areas should be treated with insecticides. Treatment must be approved by the veterinarian, since many insecticides are harmful to dogs.

Health Care: Diseases and Parasites
Parasites

Internal Parasites

Internal parasites (living in the host's body) irritate the tissues, rob the body of blood or essential elements of its diet, or interfere with specific body functions. An understanding of the life cycle of internal parasites is important to controlling and preventing infestation. The life cycles of several of the most commonly found internal parasites are listed and then discussed below.

- Heartworms
- Hookworms
- Roundworms
- Tapeworms
- Whipworms

Canine officers should consult with a veterinarian to determine the best way to prevent internal parasites.

Heartworms

Heartworms are found in the heart and lungs of a dog rather than the intestine. Heartworms are threadlike in appearance, are 6-8" long, and interfere with a dog's heart action and circulation.

FIGURE 3-2-1 shows the life cycle of a heartworm. The adult worms in the heart produce larvae called microfilaria. They circulate in the dog's bloodstream where they may be picked up by mosquitoes, the insect responsible for spreading the parasite. The larvae continue their development in the mosquito and then are injected into a dog's tissues when the mosquito feeds. The microfilaria travel to a dog's heart and develop into adults.

Symptoms of heartworms include: coughing, loss of weight, difficult breathing, and quick loss of energy. This parasite is diagnosed by a veterinarian during a blood test which reveals microfilaria in the bloodstream, if present. Monthly or daily medication is given to kill the microfilaria. Adult heartworms are killed by extensive treatment by a veterinarian. The treatment consists of several injections that kill the adult heartworm and of several months of rest for the dog while the dead heartworm is reabsorbed into the dog's system.

Control of heartworms includes: treating the infested dog to prevent them from serving as sources of the infestation and controlling mosquitoes in the area. Annual blood work detects early stages of heartworm.

Health Care: Diseases and Parasites
Parasites

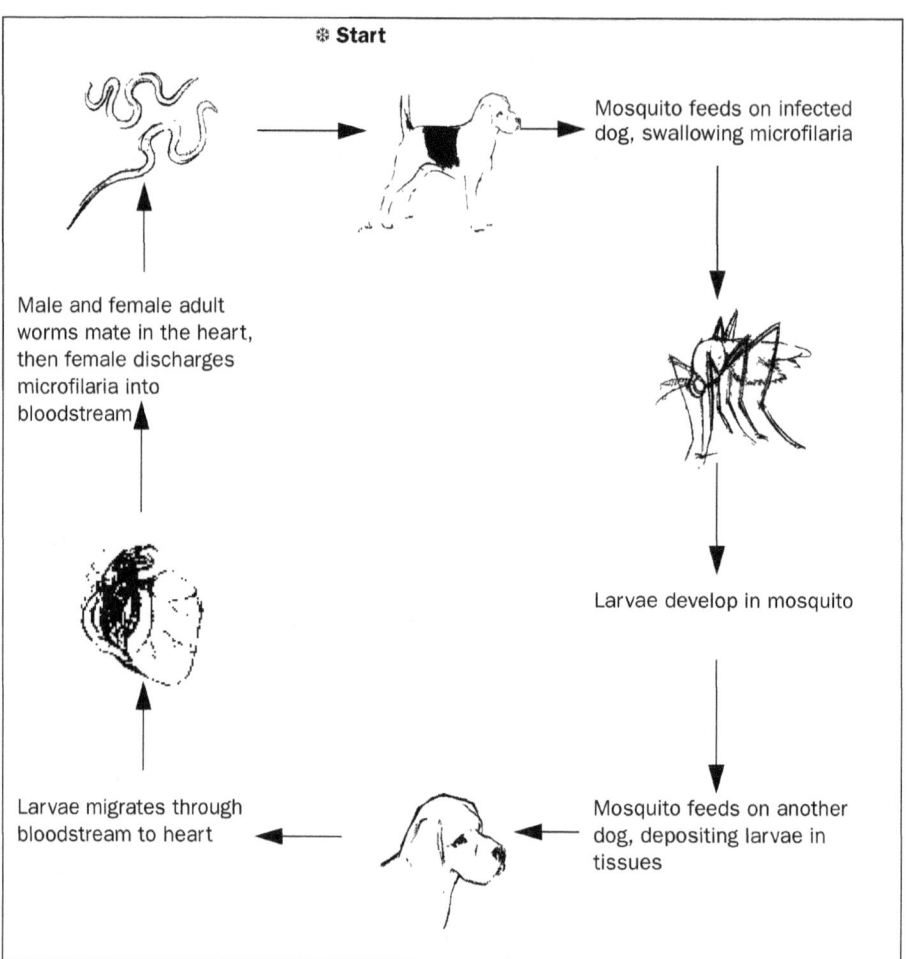

FIGURE 3-2-1: Life Cycle of a Heartworm

Hookworms

Hookworms live in a dog's intestines and are one of the most harmful parasites. They are small and thread-like (1/2–3/4" long). They suck blood and cause blood loss by tearing the intestinal wall.

FIGURE 3-2-2 shows the life cycle of a hookworm. The adult worm lives in a dog's intestine where the female produces eggs that pass through a dog's stool. Larvae develop from these eggs and can infest the same dog or another one. The larvae penetrate the skin or are swallowed as a dog licks the ground or itself. The larvae pass directly into the lungs, are coughed up and swallowed, and then reach the intestine. Once in the intestine, they develop into adult hookworms, and the cycle begins again.

Health Care: Diseases and Parasites
Parasites

Dogs with hookworms may have a variety of symptoms, depending on the severity of the infestation. A veterinarian must diagnose the disease by microscopic examination of the dog's stool. Symptoms may include: pale membranes in the mouth and eyes, loose stools containing blood, or loss of weight.

Control of hookworms includes: treating the infested dog and keeping the area free of fecal matter.

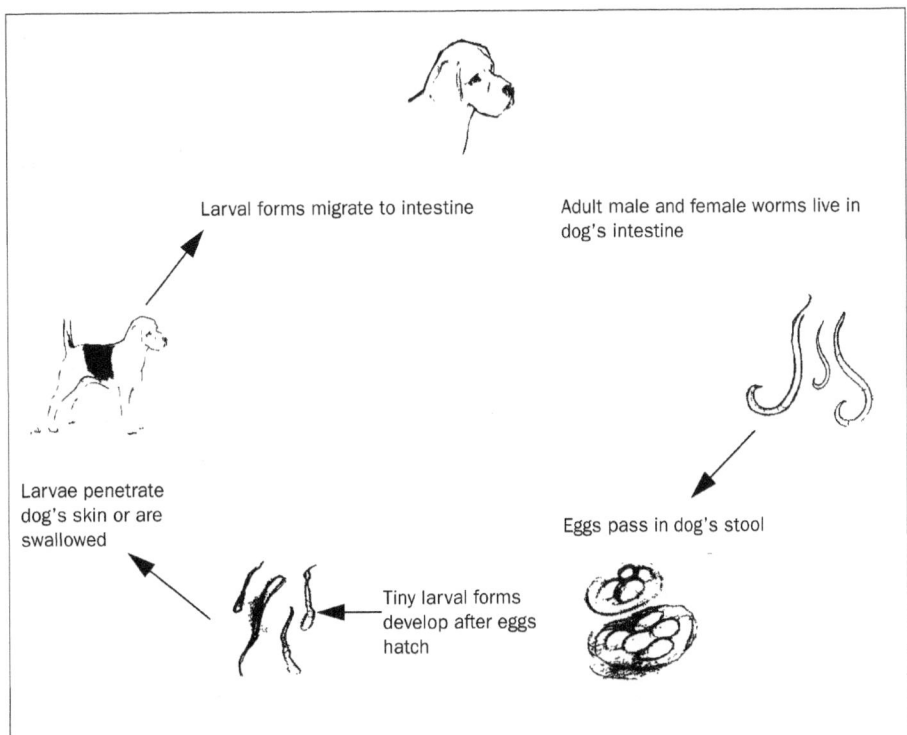

FIGURE 3-2-2: Life Cycle of a Hookworm

Roundworms Roundworms live in a dog's intestine and range from 2-8" long. The life cycle of the roundworm is similar to that of the hookworm (refer to **FIGURE 3-2-2**), except the eggs of the roundworm do not develop into larvae until they have been swallowed by a dog. Adult roundworms rob an infested dog of essential nutrients in the diet, and larvae of roundworms irritate as they travel through the lungs.

Symptoms may include: vomiting, diarrhea, loss of weight, and coughing. A canine officer can diagnose roundworms by finding eggs or adult roundworms in a dog's stool or vomit.

Control of roundworms includes: treating the infested dog and practicing good sanitation in the kennel area.

Health Care: Diseases and Parasites
Parasites

Tapeworms

Tapeworms are long, flat, and ribbon-like. They have many segments and a head. The tapeworm attaches its head to the wall of the intestine. Several kinds of tapeworms may infest a dog.

FIGURE 3-2-3 shows the life cycle of a tapeworm. It is rather complex. After eggs of a tapeworm have passed through a dog's stool, they are eaten by the larvae of a flea. The larvae of a tapeworm develops when the adult flea (or lice) is eaten by a dog. The larvae enters the dog's intestine and develops into an adult tapeworm.

The symptoms of tapeworms are usually not too noticeable but may include: diarrhea, loss of appetite, and loss of weight. Often the eggs of the tapeworm cannot be detected by a veterinarian during stool examinations. However, tapeworm segments passed by an infested dog may be seen in the stool or among the hairs in the anal region. These segments are called "crawling" proglottids and are small, white objects about 1/4" long (like small grains of white rice). The word crawling is used only when they are fresh and moving.

Control of tapeworms requires: treating the infested dog, practicing good sanitation in the kennel area, controlling fleas, and disallowing a dog to eat animal meats that are likely sources of infestation. Such animal meats include rabbits, rodents, sheep, and ungulates (hoofed animals), such as deer, swine, horse, cattle, and elephants.

Health Care: Diseases and Parasites
Parasites

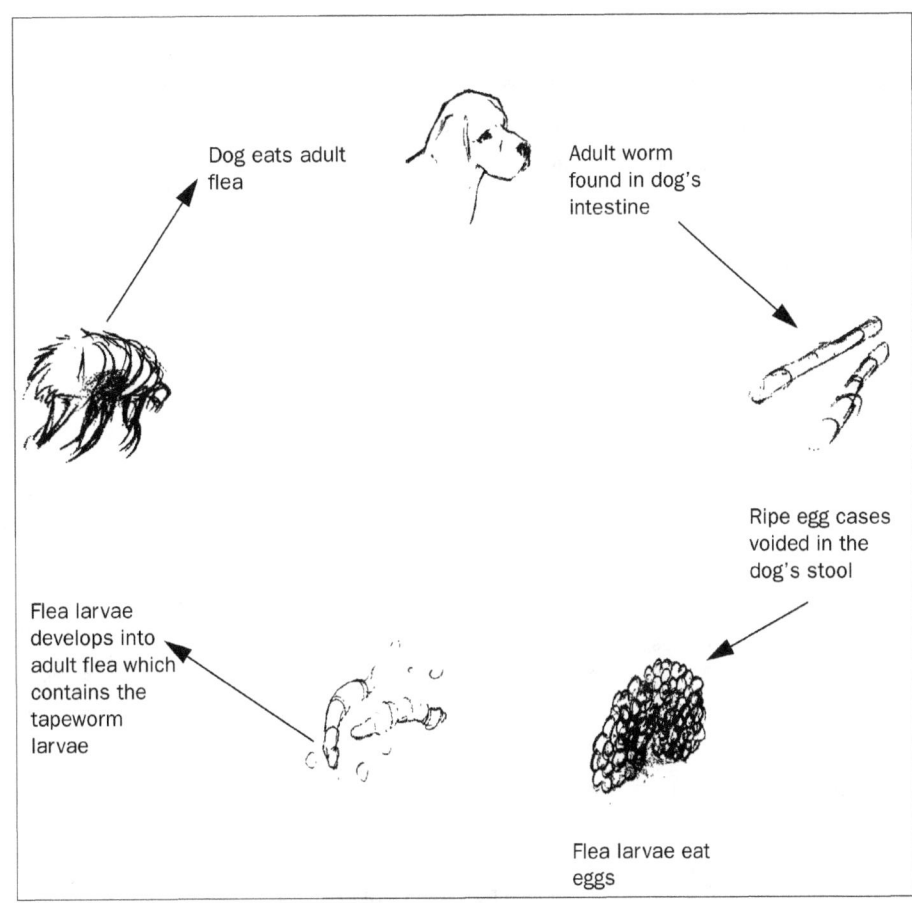

FIGURE 3-2-3: Life Cycle of a Tapeworm

Whipworms Whipworms are smaller than roundworms (2-8" long) but larger than hookworms (1/2–3/4" long). The life cycle of the whipworm is very similar to that of the roundworm, except the larvae of the whipworm do not travel to the lungs before becoming adults in the intestine.

Symptoms may include: diarrhea, loss of weight, and pale membranes of the mouth and eyes. A veterinarian must diagnose the disease by microscopic examination of a dog's stool. Control of whipworms includes: treating the infested dog and practicing good sanitation in the kennel area, the same measures as for roundworms.

Summary Refer to **Table 3-2-2** and **Table 3-2-3** for a summary of the external and internal parasites that could infest detector dogs. The tables summarize the symptoms to look for when observing dogs and kennel facilities. Canine officers should notify the veterinarian when a detector dog shows one or more of the symptoms of parasites.

Health Care: Diseases and Parasites
Parasites

Table 3-2-2: Summary of External Parasites

Parasite:	Type of irritation:	Symptoms:	Parasite/ disease carried:	Control:	Affects human:
Flea	◆ Bite ◆ Irritate skin	◆ Excessive scratching ◆ Chewing	◆ Tapeworms ◆ Bubonic Plague	◆ Good sanitation ◆ Pills ◆ Long term topicals ◆ Dips ◆ Powders ◆ Sprays	Yes
Lice	◆ Bite ◆ Suck blood	◆ Poor health ◆ Poor hair coat ◆ Small, white/gray crescent-shaped objects attached to hair	◆ Skin infections ◆ Tapeworms	◆ Good sanitation ◆ Dips ◆ Powders ◆ Sprays	Yes
Mite	◆ Ear mites ◆ Skin mites	◆ Dark, waxy discharge in ear canal ◆ Skin disease ◆ Hair loss	Ear infections	◆ Good sanitation ◆ Dips ◆ Powders ◆ Sprays	Yes
Mosquito	Suck blood	◆ Irritation ◆ Scratching	◆ Heartworms ◆ Encephalitis	Good sanitation	Yes
Tick	Suck blood	◆ Small, flat insect the size of a match head attached to the skin by the mouth parts ◆ Puffed-up insect the size of a pea attached to the skin by the mouth parts	◆ Lyme disease ◆ Rocky Mountain spotted fever ◆ Canine Ehrlichiosis ◆ Tick paralysis	◆ Well kept grounds at kennel ◆ Good sanitation ◆ Daily skin checks ◆ Dips ◆ Powders ◆ Sprays ◆ Long term topicals	Yes

Health Care: Diseases and Parasites
Parasites

Table 3-2-3: Summary of Internal Parasites

Parasite:	Method of infection:	Lives in:	Symptoms:	Method of diagnosis:	Affects human:
Heartworm	Microfilaria passed by mosquito during bloodsucking	◆ Heart ◆ Blood vessels	◆ Pneumonia ◆ Coughing ◆ Loss of weight ◆ Difficulty breathing ◆ Loss of strength and energy	Knott's or Difil test	No
Hookworm	◆ Ingests eggs ◆ Larvae penetrates skin	Intestine	◆ Loss of weight ◆ Blood in stool ◆ Pale membranes	Fecal exam	Yes
Roundworm	◆ Ingests eggs ◆ Prenatal from mother	◆ Intestine ◆ Stomach	◆ Poor hair coat ◆ Loss of weight ◆ Coughing up worms	Fecal exam	Yes
Tapeworm	Ingests infested, intermediate host (flea, lice)	Intestine	◆ Loss of weight ◆ Poor hair coat ◆ Small, white worm segments in stool	Visual-segments seen in stool or on hairs around anus	Yes
Whipworm	Ingests eggs or larvae	Intestine	◆ Loss of weight ◆ Poor hair coat	Fecal exam	Yes

Health Care: Diseases and Parasites
Parasites

Health Care

First Aid and Emergency Care

Contents

Introduction **page-3-3-1**
Physical Restraint **page-3-3-2**
Bleeding **page-3-3-3**
Bloating **page-3-3-4**
Cold Injury **page-3-3-5**
Foreign Objects in the Mouth **page-3-3-5**
Fractures **page-3-3-6**
   First Aid for Fractures **page-3-3-6**
   How to Splint a Fracture **page-3-3-6**
Overheating **page-3-3-7**
Poisoning **page-3-3-8**
   First Aid for Poisoning **page-3-3-8**
Seizures **page-3-3-10**
   First Aid for Seizures **page-3-3-10**
Shock **page-3-3-11**

Introduction

In most cases, when a Canine Officer recognizes the early signs of injury or sudden illness, it allows ample time to get help from a veterinarian. However, situations may sometimes arise when medical help is not available immediately, and the seriousness of the incident requires the Canine Officer to apply first aid to protect the life or health of the detector dog.

If you cannot find what you are looking for in this section, use **Table 3-3-1** to locate information about first aid related topics elsewhere in the manual.

TABLE 3-3-1: Location of First Aid Related Topics in the Manual

If you are looking for information about:	Then refer to the following:
Operational procedures for an injury or sudden illness	"Injury and Sudden Illness" on **page-2-2-8**
Places to contact in case of an incident	"Incident Contacts" on **page-2-2-9**
How to administer medication	"Administering Medication" on **page-3-4-23**
How to express anal glands	"Expressing Anal Glands" on **page-3-5-7**
Reverse sneezing	**Table 3-4-6 on page-3-4-15**
How to make an emergency Elizabethan collar	"Elizabethan Collar" on **page-D-1-2**
The contents of a first aid kit	"First Aid Kit" on **page-D-1-8**

Health Care: First Aid and Emergency Care
Physical Restraint

Most of the injuries and illnesses listed in this section are the most common ones encountered by working dog handlers; some are less common but extremely severe. The first aid instruction provided here is the most likely to save life, prevent further injury, and reduce pain. There are many valuable books, internet sites, and an increasing number of canine first aid seminars that provide current information about administering first aid. Also, consult your veterinarian if you wish to increase your knowledge.

The first action you should take in an emergency situation is a visual assessment of the victim. Use the "ABCS" method to remember the most important observations:

- Air: Are the airways (throat, mouth) clear? What is the respiratory rate and effort?

- Bleeding: Is there evidence of external or internal bleeding?

- Consciousness: Is the victim conscious? To what degree?

- Structural abnormalities: What is the position and movement of limbs?

In all emergency situations:

- Notify a veterinarian as soon as possible, with as much information as possible about the victim.

- Get help from anyone available.

Physical Restraint

1. Calm the dog and immobilize it. Talk to the dog in a calm voice.

2. If the dog does not respond to your voice or touch because of pain and distress, you may wish to physically restrain the dog with a short tether or muzzle. Some dogs experiencing extreme pain and fear may bite, particularly if injuries are handled roughly. However, be aware that restraining some dogs with a muzzle greatly increases their fear and anxiety and they may struggle to free themselves from the muzzle, resulting in greater injury. The handler should assess whether restraint is absolutely necessary.

3. If you must muzzle the dog, do the following:

   A. Slide one hand down the leash and grab the collar where the leash attaches to the collar. Firmly hold the collar and leash clip.

B. With your other hand, grab the leash just above where you are holding the collar and leash clip. Quickly wrap the leash around the dog's muzzle one or two times.

C. After securing the leash around the dog's muzzle, put the end of the leash in the hand holding the collar. You should be holding the end of the leash from the wrapped muzzle, the leash clip, and the collar in one hand. This will keep the dog from getting out of the muzzle and will secure its head.

D. When ready to release the dog from the muzzle, hold the end loop of the leash in your hand. Let go of the muzzle, leash and collar while standing up and stepping back. In this way you may avoid being bitten.

NEVER use a muzzle if the dog is showing any of the following symptoms:

- Overheating
- Vomiting
- Difficulty breathing
- Injury in the head or neck area
- Unconsciousness
- Shock

Bleeding

Bleeding can be external or internal (hemorrhage). Notify a veterinarian of any bleeding even if you feel it is minor, especially bleeding that results from a fight with another animal. Use **Table 3-3-2** for signs of and first aid for bleeding.

TABLE 3-3-2: Signs of and First Aid for Bleeding

If bleeding is:	Then use the following first aid:
External—Blood is flowing from an open wound; treat as external bleeding	1. Apply direct pressure to the wound (particularly wounds in the dog's foot or leg, which bleed freely). Use a sterile bandage, a clean handkerchief, or pinch the wound edges together with your fingers. 2. Apply a pressure bandage as soon as possible. 3. Wrap the dog in a blanket or coat to keep it warm. 4. Immediately take the dog to a veterinarian.
Internal—Bleeding from a body opening and/or there are signs of shock; treat as internal bleeding	1. Wrap the dog in a blanket or coat to keep it warm. If necessary, cradle the dog to your body to help provide heat. 2. Immediately take the dog to a veterinarian.

Health Care: First Aid and Emergency Care
Bloating

Bloating

Bloating occurs when the dog's stomach fills with gas that cannot be adequately expelled. The stomach swells and may rotate within the abdomen until it is twisted and flipped upside down. This prevents blood flow to the lower abdomen and can cause shock. If not treated within approximately 30 minutes, this condition will result in the dog's death. Bloating may occur if a dog is fed too much at one time, if fed immediately before or after strenuous exercise, or if a dog is allowed to drink too much water too quickly immediately before or after exercise.

The one most important practice you can put into place to prevent bloating is to feed smaller meals more frequently (split daily ration into two portions, morning and evening). See **Table 3-3-3** for other precautions to prevent bloating. See **Table 3-3-4** for signs of and first aid for bloating.

TABLE 3-3-3: How to Prevent Bloating

To prevent bloating:	Then:
Before and after work	Do not feed within 2 hours.
	This does not affect the food used to reward a dog for task performance.
During exercise, training, or work	Give water in small amounts to prevent thirst.
After training	Give a minimal amount of water in a bucket. After 1 hour, fill the bucket up and allow the dog to drink as needed.
Before and after meals	Do not exercise the dog within 1 hour before and 2 hours after meals. Limit the amount of water for dogs who drink too much after eating. Soak food in warm water for 15-30 min. before feeding. This practice allows the food to expand outside of the dog's stomach.

TABLE 3-3-4: Signs of and First Aid for Bloating

Signs of bloating:	First aid for bloating:
◆ Swollen stomach just behind the ribs, primarily on the left side. Tapping stomach produces hollow "drum like" sound. ◆ Unproductive attempts to vomit or have a bowel movement ◆ Dog looks at stomach ◆ Retching, excessive drooling, and foaming at the mouth; dry heaves ◆ Restlessness, anxiety, whining, crying, pacing ◆ Inability to get comfortable in any position	1. Get help to transport the dog to a veterinarian IMMEDIATELY. 2. Call in transit to inform the veterinarian that you have a bloat case en route and of your approximate time of arrival.

Health Care: First Aid and Emergency Care
Cold Injury

Cold Injury

Identify the type of cold injury the dog has. See **Table 3-3-5** for signs of and first aid for a cold injury. Take the dog's temperature, pulse rate, respiration rate, and weight.

TABLE 3-3-5: Signs of and First Aid for Cold Injuries

Signs of cold injuries:	First aid for cold injuries:
Hypothermia: ◆ Body temperature is below normal (less than 95°F taken rectally) ◆ Shivering ◆ Decreased pulse rate ◆ Weakness ◆ Unconsciousness ◆ Shock	1. Provide warmth by wrapping the dog in blankets or towels. If necessary, cradle the dog to your body to help provide heat. 2. Take the dog to a veterinarian.
Frostbite: ◆ Exposure to extreme cold ◆ Affected tissues (ear tips, scrotum, tail, and limbs) may be reddened, pale, or scaly	1. Provide warmth by wrapping the dog in blankets or towels. If necessary, cradle the dog to your body to help provide heat. 2. Take the dog to a veterinarian.

Foreign Objects in the Mouth

A dog may get a stick or other foreign object lodged in its mouth or throat. See **Table 3-3-6** for signs of and first aid for removing the foreign object.

TABLE 3-3-6: Signs of and First Aid for Foreign Objects in the Mouth

If the dog shows any of these signs:	And you:	Then administer this first aid:
◆ Coughing and gagging ◆ Drooling ◆ Pawing at mouth ◆ Difficulty swallowing	Can see the foreign object	1. Try to remove the object by massaging the throat upward, as opposed to retrieving it from inside the mouth. 2. Get help from anyone available, if needed. 3. Notify the veterinarian.
	Cannot see the object	Take the dog to a veterinarian.

Fractures

Fractures are one or more breaks in a bone. A fracture can be simple or compound. If the fracture is contained within the skin, it is a simple fracture. If it protrudes from the skin, it is a compound fracture.

First Aid for Fractures

1. Immediately restrain the dog to prevent further injury to the fractured area and to prevent possible injury to yourself.

 Broken bones or fractures are potentially the most serious injuries since the dog will probably continue trying to move around. The dog must be physically restrained because its movement could increase the seriousness of the injury.

2. Keep the dog quiet and warm to prevent shock.
3. Take the dog to a veterinarian.
4. Before moving a dog with a fracture, apply a splint. Apply a splint to the fractured area to immobilize it and to prevent further injury. See the following directions.

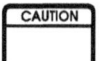 Do not try to splint fractures that are close to the dog's body.

Do not try to reset a broken bone.

How to Splint a Fracture

1. Place the limb against a rolled newspaper, piece of wood, or any stiff material at hand. Secure limb to splint with gauze, a leash, or similar material.

 If the fracture is compound (bone protrudes from the skin), then cover the area with a sterile gauze bandage before applying a splint. Take the dog to a veterinarian.

2. Fasten the splint above and below the fracture.
3. Apply the splint firmly but not so tight that the pressure stops the blood flow.
4. If you cannot splint the fracture, move the dog onto a firm platform made from strips of board or sheets of plywood large enough for the dog to lie comfortably.

Health Care: First Aid and Emergency Care
Overheating

Overheating

Overheating (hyperthermia) results when a dog is unable to eliminate body heat rapidly enough. It is caused by any of the following conditions:

- Hot external conditions (especially hot (95°F) and humid)
- Over-excitement
- Over-exertion
- Being physically unfit

To prevent overheating in hot weather, limit training and exercise and allow frequent breaks. To prevent overheating while traveling, provide adequate ventilation.

When a detector dog becomes overheated, the Canine Officer must take immediate action to save the dog's life. Refer to **Table 3-3-7** for signs of and first aid for overheating. The dog cannot return to work until released by a veterinarian.

NEVER leave a dog in an unattended vehicle in hot weather, even if the vehicle is air conditioned. Interior vehicle temperatures can rise by more than 30 degrees in 30 minutes. If for some reason the vehicle stops running, the air conditioner will stop cooling. **This may result in the dog's death!**

TABLE 3-3-7: Signs of and First Aid for Overheating (Hyperthermia)

Signs of overheating:	First aid for overheating:

◆ Heavy panting
◆ Unresponsiveness to commands
◆ Weakness or unsteady gait; unwillingness to move
◆ Vomiting, diarrhea
◆ Convulsions, seizures, and collapse
◆ Elevated body temperature (105°F or higher)
◆ Rapid pulse
◆ Bright red gums (may turn pale if the dog goes into shock)
	1. Lower the body temperature gradually:
	   a. Remove the dog from the heat, if possible (into shade, into air conditioned area).
	   b. Spray or splash lukewarm water on the dog's body and increase air flow over the dog.
	   c. Place ice packs wrapped in towels between the dog's rear legs.
	2. Take the dog to a veterinarian, even if it appears that the dog has recovered.

Do NOT submerge a heat exhausted dog in cold water or ice. This actually increases core temperature by constricting peripheral blood vessels.

Health Care: First Aid and Emergency Care
Poisoning

Poisoning

Careful control of a detector dog should prevent it from ingesting harmful items. Do not place poisonous products where the dog can find them. Poisons used for rodent and insect control around kennels should be applied only under the direction of a veterinarian.

Dogs may be poisoned by many different things in the environment. These include: chemicals, such as insecticides, cleaning solutions, and antifreeze; plants, such as seeds, bark, and leaves; and animals, such as snakes, toads, and salamanders.

The signs of poisoning vary. They may include:

- Drooling
- Vomiting
- Fatigue in combination with the above two points
- Convulsions
- Severe diarrhea soon after ingesting poison
- Witnessing consumption of poison

 Unless you are certain that a dog has eaten poison, do not treat for poisoning.

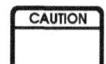 Do not attempt any first aid without contacting either the National Animal Poison Control Center (NAPCC) or the veterinarian.

First Aid for Poisoning

The following directions are provided along with other information by the National Animal Poison Control Center (NAPCC) at their web site at: http://www.napcc.aspca.org.

1. Take 30–60 seconds to safely collect and have at hand the poison involved.
2. Keep the dog quiet and warm to prevent shock.
3. **Take the dog to a veterinarian immediately.** Be sure to take with you in a plastic, zip-lock bag any product container and any material the dog may have vomited or chewed.

 Identification of the poison is the most important information for proper diagnosis and treatment.

Health Care: First Aid and Emergency Care
Poisoning

4. If you know what the dog has eaten, call local management as soon as possible and have them call NAPCC at either of the following numbers:

   ❖ 1–900–680–0000 (phone number will be billed)
   ❖ 1–800–548–2423 (have credit card ready to provide billing information)

   When the call is made to NAPCC be ready to give:

   ❖ Your name, address, and telephone number
   ❖ Credit card number for fee charged
   ❖ Information concerning the exposure—the amount of poison, the time since exposure, etc.
   ❖ Species, breed, age, sex, weight of dog; other animals involved
   ❖ Poison to which the dog was exposed
   ❖ Problems and symptoms that the dog is experiencing

Seizures

Seizures happen when there is a sudden disturbance in brain function. When a dog has a seizure, it appears to lose control of its body. Seizures may be caused by low blood sugar, liver disease, lack of oxygen, infection, poison, or brain tumors. If the seizures recur, the condition is known as epilepsy.

The warning signs depend on the severity of the seizure. If it is a mild seizure, signs may include:

- Staring blankly, acting blind
- Walking in circles, chasing the tail, "snapping at the air"
- Mild twitching
- Stiffening of muscles
- Showing behavior changes

If the seizure is more severe, the signs may include:

- Collapsing
- Twitching or shaking
- Arching the back, flailing the legs
- Trance-like state
- Urinating or defecating without apparent control
- Salivating excessively
- Rolling the eyes

First Aid for Seizures

1. **Remain calm.** Most seizures are not life threatening.

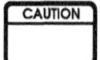 Avoid putting your fingers near the dog's mouth. Dogs rarely choke on their tongues. Do not attempt to hold your dog's mouth open or closed. Do not put anything in its mouth.

2. If your dog is having a mild seizure, try to gain its attention. This action might prevent a severe seizure from developing.

3. If a severe seizure has developed, continuing for more than a minute, clear all objects away from the dog, wrap it in a blanket, and surround it with cushions to prevent the dog from injuring itself.

4. After the seizure, keep the dog calm and confined. Unwrap the blanket from the dog. Leaving it wrapped can lead to hyperthermia (overheating).

Health Care: First Aid and Emergency Care
Shock

5. Immediately take the dog to a veterinarian.
6. Note the following about the seizure:
   ❖ The time it occurred
   ❖ Length of the seizure
   ❖ Number of hours after a meal
   ❖ What the dog was doing before the seizure
   ❖ Anything unusual that preceded the seizure

 Notify your RCPC and local management immediately regarding the seizure.

Shock

Shock is failure of the peripheral circulatory system and leads to the collapse of the cardiovascular system. Shock can be caused by bleeding, intense pain, heart failure, vomiting, diarrhea, twisted stomach, bites, poisoning, severe disease, or many other injuries, illnesses, and accidents. When there is injury to internal organs, internal bleeding may occur and the dog may go into shock. Whatever the emergency, always look for signs of shock. See **Table 3-3-8** for signs of and first aid for shock.

TABLE 3-3-8: Signs of and First Aid for Shock

Signs of shock:	First aid for shock:
◆ Pale or light pink gums ◆ Glassy look in the eyes ◆ Dilated pupils ◆ Low body temperature ◆ Cool extremities (nose, lips, feet) ◆ Rapid, shallow breathing of over 30 breaths/minute ◆ Rapid heartbeat of over 150 beats/minute ◆ Restlessness, confusion, or anxiety	1. Wrap the dog in a blanket or coat to keep it warm. If necessary, cradle the dog to your body to help provide heat. 2. Get help at once. If it is necessary to move the dog, use a litter. **3. Immediately take the dog to a veterinarian.**

Health Care: First Aid and Emergency Care
Shock

3 Health Care

General Care

National Detector Dog Manual

Contents

Introduction   **page-3-4-1**
Health Care Considerations for Working Dogs   **page-3-4-1**
Veterinary Health Checkups   **page-3-4-2**
  Semiannual   **page-3-4-2**
  Annual   **page-3-4-2**
Daily Health Checks   **page-3-4-2**
Monitoring the Health of the Kennel Environment   **page-3-4-7**
Maintaining Peak Performance through Nutrition and Conditioning   **page-3-4-7**
  Nutrition   **page-3-4-7**
  Conditioning   **page-3-4-8**
  Weight Management   **page-3-4-8**
Recognizing Symptoms of Illness   **page-3-4-9**
  Vomiting   **page-3-4-18**
  Other Abnormal Behaviors   **page-3-4-22**
Immunization Schedule   **page-3-4-22**
Microchips   **page-3-4-22**
Other Medical Considerations   **page-3-4-23**
Administering Medication   **page-3-4-23**
  Capsules, Tablets or Liquids   **page-3-4-23**
  Eye Ointment and Eye Drops   **page-3-4-25**
  Ear Drops and Ear Ointment   **page-3-4-27**
Selecting Veterinary Services   **page-3-4-28**

Introduction

Prevention is perhaps the most important element in maintaining a healthy working dog. Performing daily health checks on your dog, monitoring the condition of the kennel environment, identifying deviations from your dog's peak physical and mental condition, providing adequate nutrients, and recognizing symptoms of illness are all pivotal to providing preventive care.

Health Care Considerations for Working Dogs

Working dogs and dogs housed in kennels are under a great deal of stress. Stress stimulates the adrenal glands to produce corticosteroids, which, even at continuous low levels, are sure to compromise the dog's immune system over time. Additionally, detector dogs are exposed to an extraordinary array of pathogens in their work environment, and often, too, in the kennel environment. The kennel environment, itself, may wear on the dog's immune system through the indiscriminate use of cleaning materials or inadequate hygiene. These factors make working dogs ready targets for a variety of health problems.

Veterinary Health Checkups

Each canine must have the following veterinary checkups:

Semiannual

This is a general examination, mostly external, which should include checking the coat and skin, feet and legs, eyes, nose, ears, mouth, genital and rectal areas, vital signs (TPR[1]), and fecal tests.

Annual

This is a full examination, which includes a general examination, vaccinations, complete blood-work, teeth cleaning (if needed), x-rays, and fecal tests.

Daily Health Checks

The following table lists values that are within normal ranges for most dogs. However, each dog may vary and it is important that the Canine Officer note the normal values for his or her individual dog. Any substantial deviation from normal should be investigated.

TABLE 3-4-1: Canine Vital Signs

Function	Normal	Measure	Other	Signs of Health Problems
Respiration	10–30/min	Chest rise (count for 15 seconds x 4= respirations per minute	Panting can reach nearly 100/minute	◆ Neck extended, unwilling to lie down, labored breathing ◆ Restrictive: increased rate, decreased depth (pain, compromise of lower respiratory tract) ◆ Obstructive: normal-to-increased rate, increased depth. Obstructed airways, swelling, snake bite, neck trauma
Heart Rate	70–160/min (size of dog influences rate)	Pulse points Direct cardiac Femoral Digital	Some deviation from a normal heartbeat can be expected in athletic dogs	◆ Comparing pulse at each point may indicate where problems are occurring.
Temperature	99°F–102°F	Rectal thermometer	Aural thermometers are not accurate	◆ High: heat stroke, fever ◆ Low: unable to thermoregulate; shock, infection

---

1 TPR = **T**emperature, **P**ulse, **R**espiration Rate.

Health Care: General Care
Daily Health Checks

Canine Officers must perform daily health checks on their detector dogs to ensure good health (both physical and mental). Perform daily health checks while brushing the dog.

Each Canine Officer should know how his or her dog's coat normally looks, the frequency of its bowel movements, its eating habits, and its normal body temperature at rest. Canine Officers use this knowledge when checking their dogs to help reveal anything abnormal. Abnormalities, along with symptoms of diseases and parasites, will help detect illness in the early stages.

Refer to other sections under *Health Care* for dog anatomy, diseases, parasites, first aid, and emergency care. Knowing the proper terms to describe the dog's anatomy and to describe symptoms of injury or illness will enable Canine Officers to more efficiently report problems to their veterinarians.

Use **Table 3-4-2** that follows as a guide when checking the dog for symptoms of illness or injury. If a symptom is present, contact the veterinarian. Also, check the kennel and run areas daily (See ***Monitoring the Health of the Kennel Environment***).

Health Care: General Care
Daily Health Checks

TABLE 3-4-2: Daily Health Check

Check the dog's:	The following could be symptoms of illness or injury:
Eyes: Illnesses are frequently accompanied by changes in the eyes and many illnesses affect only the eyes. Usually, a dog's eyes are bright and clear and the surrounding membranes are a healthy pink.	◆ Red or yellow color of the membranes and white part of the eyes ◆ Paleness of the membranes ◆ White or yellow discharges ◆ Cloudiness or other discolorations of the clear part of the eyes (cornea) ◆ Puffy eyelids ◆ Eyelid partially or completely closed ◆ Nictitating membrane* that covers more of the eyeball than usual *The nictitating membrane, or third eyelid, is the small, wedge-shaped membrane at the inner corner of the eyes. Usually, this membrane covers only a small part of the eye.
Nose: The black pad at the end of a dog's nose is usually shiny and moist.	◆ The black pad is persistently dry, dull, and warm ◆ Watery, yellowish, or red-tinged discharge ◆ Sneezing ◆ Snorting ◆ Pawing at the nose
Ears: The external portion of the ear is called the flap. The vertical ear canal extends down in the earflap to the opening of the horizontal ear canal that leads to the inner ear. A small amount of brownish wax in the vertical canal is normal. (continued on next page)	◆ Reddish, discoloration of the ear canal ◆ Swelling of the ear canal ◆ Unpleasant odor coming from the ear canal ◆ Shaking of the head, holding the earflap down, holding the head to one side, twitching the ear, scratching or pawing at the ear ◆ Evidence of pain when the ear is touched ◆ Large amount of wax in the ear canal* *CAUTION: Never probe into the ear canal. You can remove dirt and wax from the inner part of the earflap. Have the veterinarian check the ears even when they appear to only need cleaning. NOTE: See **Cleaning the Ears** in the **Grooming** section.
Mouth: In the dog's mouth, gums, and inner lips should be a healthy pink. Teeth should be firm and white. Brush teeth as recommended by a veterinarian.	◆ Paleness or discoloration of gums ◆ Sores on gums ◆ Persistent drooling ◆ Bloody saliva ◆ Gagging or pawing at the mouth ◆ Bad breath, or worse than usual ◆ Loose and broken teeth ◆ Tartar accumulations on the teeth ◆ Objects lodged between the teeth or in roof of mouth

Health Care: General Care
Daily Health Checks

TABLE 3-4-2: Daily Health Check (continued)

Check the dog's:	The following could be symptoms of illness or injury:
Coat and skin: A well-fed and groomed dog usually has a glossy coat and skin that is soft and pliable. Note that the coat can change appearance with climate and season. The undercoat is thicker in cold weather and sheds in hot weather.	◆ Reddening ◆ Scabbing ◆ Moist discharges ◆ Scratching ◆ Abnormal shedding ◆ Loss of hair in spots ◆ Dryness ◆ Loss of pliability ◆ Presence of fleas or ticks
Feet and legs: Foot pads should be free of foreign objects, cuts, bruises, and abrasions. (continued on next page)	◆ Foreign objects, cuts and bruises, and abrasion of the pads ◆ Long, broken, or split nails (nails should not touch the ground when the dog stands) ◆ Loosely attached dewclaws should be removed by a veterinarian ◆ Long nails on dewclaws (nails should not curl around to the pad) ◆ Lameness ◆ Wounds, swelling, or sores on legs ◆ Inflamed elbow callus
Genital area: In a male dog, the penis is located in a fold of skin known as the prepuce or sheath. A small amount of greenish-yellow discharge at the end of the sheath is normal. In a female dog, the external genital opening is the vulva. Usually, there is no discharge.	**Males:** ◆ Large amounts of discharge present ◆ Bleeding sheath ◆ Blood in urine (after the dog has urinated, look at the end of the sheath for blood) ◆ Swelling, reddening, or scabbing of the scrotum (the pouch normally containing the testicles) ◆ Frequent unproductive attempts to urinate. Quickly licks sheath after attempting to urinate. (Signs of urinary tract infection.) **Females:** ◆ Reddening of the vulva or the skin in the genital area ◆ Discharge ◆ Blood in urine (Watch the dog urinate to detect blood. If you detect blood, note if the blood was in the first portion, last portion, or distributed throughout the urination.) ◆ Increased urination, beyond the ordinary ◆ Frequent unproductive attempts to urinate. Quickly licks vulva after attempting to urinate. (Signs of urinary tract infection.)

Health Care: General Care
Daily Health Checks

TABLE 3-4-2: Daily Health Check (continued)

Check the dog's:	The following could be symptoms of illness or injury:
Rectum area: The opening from the rectum is the anus. On either side of the rectum near the anus is a small sac (anal gland) that is a frequent source of trouble.	◆ Swelling or reddening of the skin in the area of the anus ◆ Biting at the rectum area or sliding along in a sitting position (symptom of anal glands being full or infected) ◆ Soft or watery stool ◆ Blood in the stool ◆ Worms or segments present in the stool; worm segments on hair around anus ◆ Difficulty or straining while eliminating waste
Body: The general appearance of the dog (continued on next page)	**Females**: Presence of lumps on the dog's body. Take special care around teats (checking for breast cancer).
Attitude and actions: The dog's attitude and actions are best indications of general health.	◆ Undue nervousness ◆ Loss of vitality and energy ◆ Increased desire for sleep ◆ Tiredness ◆ Inattentive while working or training ◆ Changes in appetite, thirst, or breathing ◆ Vomiting or blood in the vomit ◆ Stressed (panting excessively or circling the kennel run) ◆ Any suspicious behavior that is not typical for the specific dog
Temperature: A dog's temperature is also an excellent indication of the animal's health. Know your dog's normal body temperature while at rest. Usually, a dog's normal body temperature is within 101–102 °F.	Unusual variation in temperature. Some variation in temperature may be normal, such as following exercise. Take the temperature rectally following these directions: 1. Depending on your dog's previous behavior, you may have to muzzle the dog before taking its temperature. 2. Lubricate the rectal thermometer with petroleum jelly to ease insertion. 3. Insert the thermometer only 1" into the rectum. 4. Hold the thermometer in the rectum for 2–3 min. before reading. Hold the thermometer in place while taking the temperature to prevent the thermometer from completely entering the rectum.
Hydration: When vomiting and/or diarrhea persist, a dog can become dehydrated rapidly. Also, dehydration can be a sign of an underlying illness. Note that it may be normal for a dog to eat grass and vomit one time.	◆ Skin lacks turgor or pliability (when gently pulling and releasing the skin above the shoulders, it should immediately return to normal) ◆ Dry mucous membranes ◆ Eyes appear sunken back in their orbits ◆ Slow capillary refill time—more than 2 seconds ◆ Rapid heartbeat of over 150 beats/minute ◆ Slow heartbeat of under 80 beats/minute **CAUTION:** Dehydration is a serious condition; when symptoms are present, immediately take the dog to the veterinarian.

Monitoring the Health of the Kennel Environment

Kennel requirements are discussed in detail in Section 2-3-2. Daily kennel checks should be performed for the presence of any abnormal conditions such as blood, vomit, worms, or insect pests. The kennel should be clean and dry whenever the dog is present. Kennel runs must be clear of insect pests, reptiles, rodents, or any other organism that can transmit a disease or parasite. Additionally, the kennel run should be free of debris, physical hazards such as loose wires, sharp metal points or edges, and other items that could injure the dog. Kennel cleaning solutions should not be irritating to the dog. Vaporous irritants, such as gas from chlorine bleach, can severely stress your dog, as well as cause irreparable damage to fragile respiratory tissues.

At the beginning of your tour of duty, check the kennel for:

- Stools (watery or runny, mucous, tapeworms, abnormal color)
- Evidence of blood stains, vomit, abnormal urine stains
- Crawling insects (ticks)
- Food and water (availability and freshness). Note the amounts consumed by the dog.

At the end of your tour of duty, check the kennel for:

- Safety hazards (broken fences and/or unworkable locks or handles)
- Bucket or pan full of fresh water. If there is an automatic water supply system, then make sure it is working.
- General cleanliness of the kennel run

Maintaining Peak Performance through Nutrition and Conditioning

Nutrition

What Food to Use

Premium pet foods are required. They can be purchased at pet stores, veterinarian offices, and specialty stores.

When to Feed

Dogs must be fed at least once in a 24-hour period. Exceptions are when a feeding schedule is prescribed by a veterinarian for medical purposes. Canine Officers should maintain a routine feeding schedule.

Develop and Maintain a Feeding Schedule

At work locations where there is more than one detector dog team, Canine Officers must coordinate with each other and the kennel personnel when developing and maintaining feeding schedules.

Conditioning

As endurance athletes, detector dogs should be conditioned gradually into increased levels of exertion. Allowing your dog to progress in physical activities at a reasonable pace ensures that the dog stays enthusiastic about the activity, and also minimizes the potential for stress injury. Conditioning for peak performance should include a warm-up, exertion phase, and cool down. Massaging muscles that have been stressed is relaxing to the dog and helps build rapport between dog and handler. Massage is also an excellent way to locate any sore areas, compromised muscles, or stiff joints.

To maintain its health and welfare, the dog must have a scheduled 24-hour day off each week.

Weight Management

Obesity (actual body weight exceeds ideal weight by 15-20%) is one of the most common health problems for dogs. Obesity should be considered a serious health problem, as it can increase a dog's disposition for skin problems, heart failure, hypertension, and orthopedic problems. Obesity suppresses the immune response, increases the risks associated with anesthesia, and inhibits recovery from surgery. It is typically caused by providing the dog more energy (food) than it needs or uses. Other factors that may lead to obesity in canines include: neutering (tends to increase appetite, especially in females), genetics (especially beagles, Labrador retrievers, among other breeds), free-feeding of high quality food, competition for food, obesity as puppy.

Assessment of proper body weight is simple. The presence of a moderately defined "waist" when viewed from above, or abdominal "tuck" when viewed from the side indicates that the dog is not obese (refer to **Appendix F**.) Light palpation of the ribcage is another indicator of weight condition. Ribs should not be visible, but should be easily palpable without applying pressure in a dog in good condition. If you must apply pressure to feel the ribs, your dog is probably carrying extra weight.

Health Care: General Care
Recognizing Symptoms of Illness

Canines can lose 2-4% of their body weight per week without ill effect. Actual hunger can be detrimental to the dog's working ability, as it may cause the dog to lack focus or become irritable.

Do not feed supplements without veterinary advice. Supplements can unbalance a balanced diet and cause problems with joints and other organ systems. Do NOT feed supplements such as bones or meat. These items can create digestive upsets and other, more serious problems.

Recognizing Symptoms of Illness

Canine Officers should be aware of symptoms of illness that could have a negative effect on the working proficiency of their dogs. Veterinary medicine has vaccines for most of the deadliest infectious diseases that effect dogs (please see addendum at the end of the chapter for descriptions of these), however dogs are subject to other ailments that require the attention of a veterinarian. At minimum, a Canine Officer should be aware of symptoms that probably signal that the dog needs the services of a veterinarian. Early detection and treatment of an illness may limit its severity or may prevent a more serious condition from developing.

The tables below list several common symptoms Canine Officers may encounter, and assist the Canine Officer in determining if the condition warrants a veterinary consultation. Daily health checks are the best way to detect symptoms early. Daily health checks will also help each Canine Officer calibrate the "normal" for his or her dog, so that deviations from normal will be apparent. At the first sign of any abnormality a Canine Officer should begin documenting the condition. The Canine Officer can greatly enhance the effectiveness of diagnosis and treatment by providing certain types of information to the veterinarian.

Following are checklists of information that may be helpful to the Canine Officer for organizing information for the veterinarian. These lists may be photocopied for future use.

Health Care: General Care
Recognizing Symptoms of Illness

TABLE 3-4-3: Health Abnormality Checklist

What is the abnormality? (Describe, do NOT Prescribe)	
When did you first notice it?	
Does the abnormality occur at any **particular time** (e.g., after meals? At night?)	
How many times has it occurred?	
Did **any unusual events or situations** precede the condition?	
Has anything **changed in the dog's environment**?	
Does the condition occur in **any particular setting**?	

TABLE 3-4-4: Symptom Checklist

Attitude	❐ Normal	❐ Depressed ❐ Lethargic ❐ Restless	Observations and Notes
Appetite	❐ Normal	❐ Increased ❐ Decreased ❐ Unnatural	
Thirst	❐ Normal	❐ Increased ❐ Decreased	
Bowels	❐ Normal	❐ Soft, formed ❐ Soft, unformed ❐ Watery ❐ Other	
Urination	❐ Normal	❐ Increased ❐ Decreased	
Breathing	❐ Normal	❐ Labored ❐ Wheezy	
Coughing	❐ No	❐ Dry ❐ Wet ❐ Hacking ❐ Retching	
Vomiting	❐ No	Frequency _____ Duration _____	

Health Care: General Care
Recognizing Symptoms of Illness

Below are tables that provide general guidelines on how to determine a course of action if you observe abnormalities in your dog. Some of the most common problem areas that may present symptoms of health problems in detector dogs include the following:

- Appetite (sudden changes in appetite either with or without weight changes)
- Coughing
- Bowel Problems such as diarrhea
- Vomiting

Eyes and ears can usually be kept fairly clean and dry with regular attention. Any odors or dark colored discharges from these areas should be referred to a veterinarian for diagnosis and treatment.

The most important determinant of whether you should call or take your dog to a veterinarian, however, is what you observe as an abnormal behavior or condition for your dog. Early signs of illness may be subtle in many dogs. If you feel your dog is "off", but is not presenting discrete symptoms, increase your watchfulness of the dog for at least 24–48 hours.

Health Care: General Care
Recognizing Symptoms of Illness

TABLE 3-4-5: Determining Action to Take on Sudden Changes in Your Dog's Appetite

If your dog's appetite:	And you also observe:	The cause MAY be:	Action	
			If:	Then:
Increases	◆ Weight gain ◆ Normal Elimination	◆ Change in diet	→	Decrease amount of food by 1/4 cup increments, weekly, to control weight.
	◆ Increased water consumption ◆ Hair loss ◆ "Potbellied" appearance ◆ Poor wound healing	◆ Cushing's Disease ◆ Pancreatic Disease	→	Call veterinarian to request laboratory tests.
	◆ Weight loss ◆ Increased water consumption ◆ Breath smelling of "nail polish" ◆ Vomiting	◆ Diabetes		

Health Care: General Care
Recognizing Symptoms of Illness

TABLE 3-4-5: Determining Action to Take on Sudden Changes in Your Dog's Appetite (continued)

If your dog's appetite:	And you also observe:	The cause MAY be:	Action If:	Action Then:
Decreases	◆ No vomiting or other symptoms	◆ Change in diet	Weight loss is undesired	Return to previous diet
			Weight loss is desired	Make sure weight loss does not exceed 2–4% of body wt./week
	◆ Excessive yawning ◆ Stereotyped scratching ◆ Whole body shaking ◆ Excessive licking or self-mutilation	◆ Stress	→	1. Try to remove or reduce any stressors from dog's environment. 2. Examine whether dog's social needs are being met. 3. Provide increased amount of exercise. 4. Provide stress-relief toys.
	◆ Drooling ◆ Excessive salivating ◆ Scratching face with paws ◆ Foul-smelling breath	◆ Dental problems ◆ Poisoning ◆ Obstruction	→	1. Examine teeth for signs of infection or abscess. 2. If you suspect poisoning, try to determine type and call poison control (888-426-4435). 3. Call a veterinarian.

Health Care: General Care
Recognizing Symptoms of Illness

TABLE 3-4-5: Determining Action to Take on Sudden Changes in Your Dog's Appetite (continued)

If your dog's appetite:	And you also observe:	The cause MAY be:	Action	
			If:	Then:
Decreases	◆ Lethargy ◆ Excessive or unnatural fatigue	◆ Fever, unknown origin	Temperature is over 104°F	Call veterinarian immediately.
			Temperature is within normal range	Watch closely for 12–24 hours and discuss with veterinarian
	◆ Drooling ◆ Vomiting ◆ Burping ◆ Lethargy	◆ Gastrointestinal upset, unknown origin ◆ Unnatural food ◆ Food allergies	Temperature is over 104°F	Call veterinarian immediately
			Temperature is within normal range	Watch closely for 12–24 hours and discuss with veterinarian
Is unnatural	The dog is eating ◆ Feces ◆ Stones ◆ Dirt ◆ Other non-food items	◆ Boredom ◆ Nutritional stress	⟶	1. Enrich kennel environment. 2. Provide stress-relief toys. 3. Consult veterinarian concerning diet. 4. If dog is eating feces, add formula containing papain to feed (e.g., "Forbid," "Deter," Adolph's meat tenderizer).
Is other than above	◆ Various signs	◆ Other	⟶	If in doubt, call your veterinarian!

Important

Whenever you are in doubt about your dog's health, call your veterinarian.

Health Care: General Care
Recognizing Symptoms of Illness

TABLE 3-4-6: Determining Action to Take on Coughing (Including Wheezing, Sneezing, and "Reverse Sneezing")

If your dog's cough is:	And is accompanied by:	The cause MAY be:	Action: If:	Action: Then:
Dry, persistent	◆ Labored breathing ◆ Retching	Internal parasites	→	Contact a veterinarian for laboratory tests
Dry, occurs after a meal or chewing on rawhide or other object	◆ Refusal to eat ◆ Restlessness ◆ Excessive saliva	Obstruction	Cough persists for 5 minutes or more	Call a veterinarian
			Cough lasts less than 5 minutes	Watch dog carefully for recurrence or other signs of illness
Dry, harsh, high-pitched (may be activated by gently compressing front of dog's throat)	◆ Sudden outburst of hacking, forceful coughs ◆ Retching ◆ Foamy mucous ◆ Gagging at end of coughing bout	Kennel cough	→	1. Quarantine dog **immediately!** 2. Contact veterinarian for diagnosis. **Note**: Even if dog has been immunized for common *Bordatella* sp., it may be infected by different or resistant species.
Painful (may be activated by tapping dog's chest)	◆ Weak cough ◆ Labored breathing ◆ Nasal discharge ◆ Depression ◆ Fever ◆ Stretched neck ◆ Inability to lie down	Pneumonia	→	Contact veterinarian **immediately!**
Low-pitched cough, more common at night or in morning	◆ Mucous ◆ White or bubbly sputum ◆ Weakness ◆ Dog may stand with legs spread, head down, while coughing	◆ Heart disease ◆ Congestion due to liquid in chest	Cough persists for 24 hours or more	Call a veterinarian.
			Cough lasts less than 24 hours	Watch dog carefully for other signs of illness.

Health Care: General Care
Recognizing Symptoms of Illness

TABLE 3-4-6: Determining Action to Take on Coughing (Including Wheezing, Sneezing, and "Reverse Sneezing") (continued)

If your dog's cough is:	And is accompanied by:	The cause MAY be:	Action: If:	Action: Then:
Wet	◆ Long periods of coughing ◆ Whistling noises while breathing ◆ Fever may or may not be present	◆ Allergies ◆ Bronchitis ◆ Obstruction	Fever is present for 24 hours or more	Call a veterinarian
			Fever lasts less than 24 hours, but the cough persists for 24 hours or more	Call a veterinarian
			Fever lasts less than 24 hours, and the cough lasts less than 24 hours	Watch dog carefully for recurrence or other signs of illness
Wheezing	Evidence that the dog has been bitten by an insect or exposed to fumes	◆ Bronchioconstriction ◆ Allergic reaction ◆ Congestive heart failure	→	Take dog to a veterinarian **immediately**!
Sneezing (chronic)	◆ Noisy breathing ◆ Nasal discharge with foul odor ◆ Bloody nasal discharge ◆ Plugged nostrils	◆ Rhinitis ◆ Bacterial infection	Cough persists for 24 hours or more	Call a veterinarian
			Cough lasts less than 24 hours	Watch dog carefully for other signs of illness

Health Care: General Care
Recognizing Symptoms of Illness

TABLE 3-4-6: Determining Action to Take on Coughing (Including Wheezing, Sneezing, and "Reverse Sneezing") (continued)

If your dog's cough is:	And is accompanied by:	The cause MAY be:	Action: If:	Then:
Reverse sneezing	◆ Loud, snorting noise through nose ◆ Attempts to "clear back of throat"	◆ Allergies ◆ Post-nasal drip ◆ Sensitive throat ◆ Compression of throat	→	1. Usually no further action needed, but vet may check dog's throat to rule out trauma. 2. Cover dog's nose and mouth, speak calmly, rub its neck, release immediately on cessation of distress.
Is other than above	◆ Various signs	◆ Other	→	**If in doubt, call your veterinarian!**

Important

Whenever you are in doubt about your dog's health, call your veterinarian.

TABLE 3-4-7: Diarrhea and Bowel Movement Irregularities

Area Affected	Consistency	Color	Frequency	Other Symptoms
Small bowel	Watery (rapid evacuation)	Yellow or greenish	3–4 times per day	Weight loss
	Pasty	Light		
	Greasy, large, rancid-smelling	Gray	3–4 times per day	Vomiting (except colitis)
	Tarry	Black		
Large bowel	Mucousy	Bright blood	Several, small, per hour	Normal appetite, No weight loss
Pancreas	Dry, pasty	White	→	Vomiting, diarrhea

Health Care: General Care
Recognizing Symptoms of Illness

TABLE 3-4-8: Determining Action to Take for Diarrhea and Bowel Movement Irregularities

If:	And:	And:	The cause MAY be:	Action:
◆ The stool is very loose ◆ The stool contains bright blood ◆ The stool contains slimy mucous ◆ Diarrhea alternates with constipation	→	The dog crouches forward or strains	◆ Colitis ◆ Stress-induced ◆ Growths in intestinal tract	Contact veterinarian for assessment and possible diet adjustment
The stool: ◆ Is loose to watery ◆ Contains no blood ◆ Contains no mucous ◆ Is foamy	The condition lasts 24 hours or longer	You observe no remarkable symptoms	◆ Gastroenteritis ◆ Parasites ◆ Bacterial infection	Consult a veterinarian
	The condition lasts less than 24 hours	You observe no remarkable symptoms	Diet inconsistency	Allow digestive system to rest by withholding food or feeding bland diet for 24 hours
◆ The stool is loose to watery ◆ Loose stools alternate with solid	The condition's length is variable, but chronic	◆ The stool often contains blood or mucous ◆ Bowel movements may be explosive	*Giardia* sp. (a protozoan)	Request testing by a veterinarian
You observe no bowel movement, but thick mucous is present in dog's rectum	The dog exhibits: ◆ Vomiting ◆ Lack of appetite ◆ Lethargy ◆ Fever	The dog attempts to defecate in a hunched-up stance	Blockage of bowel	Contact a veterinarian **immediately**

Vomiting

Vomiting is one of the most common occurrences in dogs. Vomiting and regurgitation are often confused. Vomiting is the forcible expulsion of partially digested matter; regurgitation is the passive expulsion of undigested matter. Chronic regurgitation is the most common symptom of a condition known as megaesophagus. This condition must be carefully managed under veterinary supervision.

Health Care: General Care
Recognizing Symptoms of Illness

Dogs often vomit after eating grass or other items. This is not usually any cause for concern. Other causes of vomiting are more serious and require veterinary attention.

- Acute vomiting is generally self-limiting, may be a single event and may be caused by overeating, eating unusual items (grass, feces, dirt, etc.), obstructions (such as foreign objects), chemicals (such as poisons or drugs), or infections (e.g., bacterial, viral).

- Chronic vomiting is the recurrence over hours, days, or other time periods and is generally a metabolic (such as Addison's disease), an idiopathic (such as epilepsy) or an inflammatory (such as irritable bowel syndrome) problem.

Health Care: General Care
Recognizing Symptoms of Illness

TABLE 3-4-9: Determining Action to Take for Vomiting

If vomiting is:	And you observe:	And you also observe:	The cause MAY be:	Action	
				If:	Then:
Acute (one to several times)	◆ Depression ◆ Decreased appetite ◆ Dehydration ◆ Drooling ◆ Whining ◆ Trembling	⟶	◆ Spoiled food ◆ Grass eating ◆ Over-eating	Fever is absent	1. Withhold food for 24 hours 2. Watch dog for other signs of illness 3. Perform skin turgor test
				Fever is present	1. Note what the dog vomits 2. Consult with veterinarian if vomiting continues
		◆ Yawning ◆ Shaking ◆ Pacing ◆ Eye-rolling ◆ Diarrhea	Stress	Fever is absent	1. Withhold food for 24 hours 2. Watch dog for other signs of illness 3. Perform skin turgor test
				Fever is present	1. Note what the dog vomits 2. Consult with veterinarian if vomiting continues
		Vomitus contains blood or mucous	◆ Gastro-intestinal disease ◆ Foreign object ◆ Other, serious conditions	⟶	Call veterinarian **immediately!**
Acute (projectile)	◆ Difficulty defecating ◆ Abdominal distress ◆ Possible respiratory distress	⟶	Blockage of upper gastrointestinal tract	⟶	

Health Care: General Care
Recognizing Symptoms of Illness

TABLE 3-4-9: Determining Action to Take for Vomiting (continued)

If vomiting is:	And you observe:	And you also observe:	The cause MAY be:	Action If:	Action Then:
Chronic	◆ Depression ◆ Dehydration ◆ Drooling ◆ Whining ◆ Trembling	◆ Poor appetite ◆ Listlessness ◆ Vomiting not associated with meals ◆ No continuous occurrence of symptoms	◆ Internal disorder (organ involvement) ◆ Diabetes ◆ Heavy parasite load	→	Call veterinarian for checkup and laboratory work
Dry heaves	◆ Restlessness ◆ Excessive drooling ◆ Depression ◆ Shock	Hard, distended abdomen	Gastric dilatation volvulus (torsion, bloat)	→	1. Take dog to veterinarian **immediately**. 2. Administer first aid en route (see). You have approximately 20 minutes from the onset of the mild symptoms before this condition results in death!
Other	Various symptoms	→	Other	→	**If in doubt, call your veterinarian!**

Important

Whenever you are in doubt about your dog's health, call your veterinarian.

For a single incident of vomiting, you may want to rest the dog's digestive system by feeding a bland diet. Basically, a good bland diet consists of a low or no fat protein and a carbohydrate. You may wish to mix one part low fat protein (e.g., cottage cheese, ricotta cheese, tuna in water, boiled chicken without broth) to two parts grain (e.g., rice, macaroni). Using rice as the grain portion has the added advantage of firming up the stool, if diarrhea accompanies the vomiting. Keep the dog hydrated, as vomiting can quickly deplete body fluids.

Health Care: General Care
Immunization Schedule

Other Abnormal Behaviors

Other abnormal behaviors that may indicate health-related problems include the following:

- "Scooting"—dragging the anus across the floor—indicates that the dog may have impacted anal glands.

- Drooling—although this can be associated with periodontal disease and some other mouth problems, it is often an indicator of stress. Dogs on food reward drool, sometimes copiously, when awaiting the reward on a target. It is important to calibrate what is "normal" for your dog and what is "abnormal" to determiner if drooling is excessive.

Immunization Schedule

Usually immunization schedules are set up by your veterinarian to correspond with local requirements or conditions that exist in your area. Please bear in mind that the distemper vaccine (killed virus) may cause a temporary lack of olfactory sensitivity in your dog. Your dog should probably be rested for 24 hours after the vaccines are administered. Leptospirosis vaccines should be of the type that will protect against as many strains of the disease as possible. At the present time, standard vaccines protect for 2 strains.

Giardia vaccines are recommended for dogs being kenneled. Lyme disease (*Borrelia*) vaccines should be administered in areas that are at high risk for this disease. When vaccinating, handlers should consider all areas where the dogs will be used, not simply the location of the kennel and airport at which they are normally assigned.

Currently, the following vaccines are administered on a yearly schedule: distemper, hepatitis, adenovirus, parainfluenza, leptospirosis, parvo, and corona. Bordatella (intranasal) vaccine is administered every six months. Bordatella vaccination may affect the dog's sense of smell.

Microchips

Microchips that identify your dog as a USDA detector dog are inserted at the training center. Please provide the information concerning the inserted chip to your local veterinarian.

Health Care: General Care
Other Medical Considerations

Other Medical Considerations

Some drugs may cause a temporary decrease in olfactory sensitivity for all or some odors. Drugs currently known to have this effect are steroids, some antibiotics, and killed virus distemper vaccines.

Administering Medication

There will be times when a veterinarian prescribes preventive medicine or medication for a sick or injured dog. Canine Officers are responsible for administering all types of medication to their dogs. The medicine may be a capsule or tablet, liquid, ointment, or drops.

Following are directions for administering different types of medicine. For related first aid and emergency care, refer to the information behind the green tab.

Capsules, Tablets or Liquids

When any foreign substance is placed directly into a dog's mouth, its first reflex is to spit it out. To successfully administer oral medication the dog is forced to swallow. Administering medicine is best done quickly and smoothly to keep the dog from being apprehensive and resentful. Before administering oral medicine without food, review the following steps for administering capsules, tablets, or liquids.

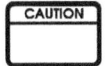 Do not give oral medications or any liquids if the dog is unconscious or cannot swallow.

Steps for Administering Capsules or Tablets

Some capsules or tablets can be put in a spoonful of canned dog food. The dog will eat the pill along with the food. Use the following steps to administer capsules or tablets without food.

1. Place the fingers of one hand over the dog's muzzle. Refer to *Figure 3-4-1*.

FIGURE 3-4-1: Placement of Hand Over the Dog's Muzzle

Health Care: General Care
Administering Medication

2. Insert thumb just behind the dog's teeth.

3. Press thumb against the roof of the dog's mouth and fingers over the dog's lips against its teeth to open its mouth.

4. With other hand, place the capsule or tablet in the center of the tongue near the back. Refer to **Figure 3-4-2**.

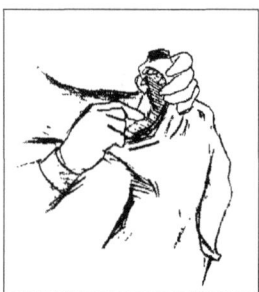

FIGURE 3-4-2: How to Administer Capsules or Tablets

5. QUICKLY remove the hand and the dog will close its mouth. HOLD the mouth closed, pointing its nose upward, while GENTLY stroking the dog's throat. Refer to **Figure 3-4-3**. Or, blow a puff of air into the dog's nose to cause a reflex to swallow.

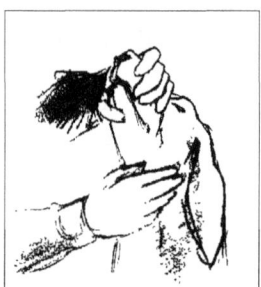

FIGURE 3-4-3: Stroking the Dog's Throat to Facilitate Swallowing Medicine

Steps for Administering Liquids

 Use extreme caution when giving an oily liquid.

1. Pour the prescribed dosage in a medicine dropper or syringe.

2. Use one hand to hold together the upper and lower jaws.

3. Point the dog's nose slightly above the horizontal. If it is raised too high, the dog won't be able to swallow.

4. Make a pouch between the molar teeth and the cheek while sliding the dropper under the dog's lip toward the back corner of the mouth. Refer to **Figure 3-4-4**. Squirt in the medicine slowly allowing the dog time to swallow the medicine.

Health Care: General Care
Administering Medication

FIGURE 3-4-4: Positioning of a Medicine Dropper or Syringe to Administer Liquids

5. If needed, rub the dog's throat. If the dog shows signs of distress (e.g., coughing or struggling), allow the dog to lower its head to rest before continuing.

Eye Ointment and Eye Drops

Before administering eye medicine, do the following:

1. Position and restrain the dog in a lying or sitting position.

2. Inspect the eyes for extraneous debris and clean the eyes, as necessary.

    A. Remove dry debris with a sterile pad or cotton ball dampened with distilled or cold water.

    B. Flush discharge from the eyes with distilled water.

3. Review the following steps for administering eye ointment or drops. Always approach the eye from behind to avoid alarming the dog.

After administering eye medicine, observe the dog for adverse reactions. If the dog rubs the affected eye with either its feet or inanimate objects, place an Elizabethan collar on the dog until the irritation subsides. Refer to **Appendix D**, **Equipment**, for a description of an Elizabethan collar and directions on how to make an emergency collar.

Steps for Administering Eye Ointment

1. Place one hand under the dog's jaw, with the thumb of the same hand on the lower eyelid. Pull the lower eyelid downward to expose the inner corner of the lower eyelid. Refer to **Figure 3-4-5**.

Health Care: General Care
Administering Medication

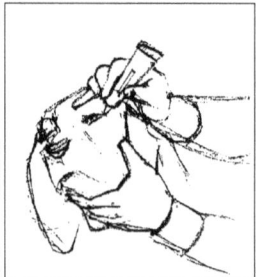

FIGURE 3-4-5: How to Hold Dog's Head While Administering Eye Ointment

2. Place the hand holding the medication on top of the dog's head with the tube directly above the surface of the eye.

3. Lay a single ribbon of ointment 1/4–1/2" directly on the inside of the lower eyelid, going from the inner to the outer part. Refer to *Figure 3-4-6*.

Important

If the ointment is thick, you may warm it under water or in your hands to make the ointment flow more easily.

CAUTION

Do not allow the tip of the tube to touch the eye or any other surface. This prevents accidental contamination of the medication or damage to the eye.

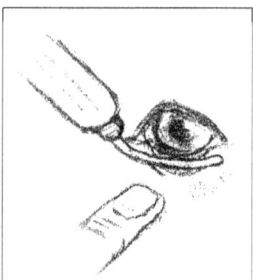

FIGURE 3-4-6: How to Apply Eye Ointment

4. Release the eyelid.

5. Repeat the procedures for the other eye, if needed.

Steps for Administering Eye Drops

1. Lift the nose upward, with one hand holding the muzzle.

2. Use the thumb of the hand holding the medication to spread the eyelids apart by pressing gently above the upper eyelid.

3. Position the bottle or dropper over the eye.

4. Drop the prescribed number of drops into the eye.

Health Care: General Care
Administering Medication

5. Release the eyelid. To ensure the drops stay in the eye, gently press.

6. Repeat the procedure in the other eye, if needed.

Ear Drops and Ear Ointment

Review the following steps for administering ear drops and ointment. Sometimes a veterinarian will have you clean the dog's ears before applying the medicine. Refer to *"Cleaning the Ears"* on ***page-3-5-5***.

Steps for Administering Ear Drops and Ear Ointment

1. Position and restrain the dog.

2. Pull the earflap upward to expose the external opening of the ear canal.

3. Administer the medicine. Refer to ***Figure 3-4-7***.

   A. Position the container directly above the external opening without the container touching any portion of the ear.

   B. Drop the prescribed number of drops or apply the ointment directly into the ear canal.

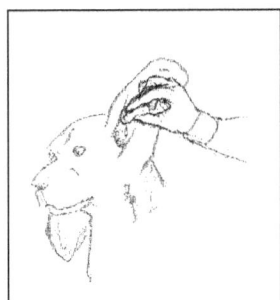

FIGURE 3-4-7: Applying Medicine to Ears

4. Massage the vertical canal with the thumb and index finger. Refer to ***Figure 3-4-8***.

FIGURE 3-4-8: Massaging Ears

5. Release the earflap of the ear.

6. Repeat the procedure in the other ear, if needed.

Health Care: General Care
Selecting Veterinary Services

Selecting Veterinary Services

Canine Officers, the RCPC, and the Port Director are responsible for selecting a veterinarian.

When selecting a health care provider for the detector dog, avoid APHIS veterinarians who have private practices. Also, quality of care to the detector dog should take precedence over fees charged.

The veterinarian must meet the following requirements:

- Have at least 50 percent of his or her practice in small animal care.

- Have been in practice for at least 1 year.

- Be accredited and licensed in the State where he or she practices. (Note that military veterinarians are accredited for domestic and international travel.)

- Have no valid complaint against him or her for animal abuse or professional misconduct. Have no violation with APHIS under the AWA. Check with the Better Business Bureau and your RCPC.

- Provide 24-hour emergency animal care. This Includes veterinarians that refer patients to emergency clinics after office hours.

- Have adequate and sanitary facilities to provide routine and emergency care. Routine care includes minor surgery, x-rays, and dental care.

Health Care

Grooming

Contents

Introduction **page-3-5-1**
Daily Brushing **page-3-5-1**
   Rubber Curry Brush **page-3-5-2**
   Rubber Slicker Brush **page-3-5-2**
   Natural Bristle Brush **page-3-5-2**
   Hound Glove **page-3-5-2**
Other Grooming Tools **page-3-5-2**
   Wire Slicker Brush **page-3-5-2**
   Undercoat Comb **page-3-5-2**
   Mat Splitters (combs, rakes, blades) **page-3-5-2**
Conditioning the Coat **page-3-5-3**
Trimming the Nails **page-3-5-3**
Cleaning the Ears **page-3-5-5**
Cleaning the Eyes **page-3-5-6**
Cleaning the Teeth **page-3-5-6**
Expressing Anal Glands **page-3-5-7**
Bathing **page-3-5-7**

Introduction

Grooming is an important daily task and often incorporates the *"Daily Health Checks"* on **page-3-4-2**. Grooming is critical for maintaining the cleanliness and professional appearance of the dog, as well as maintaining its safety, comfort, and health. A complete grooming consists of bathing the dog, brushing the coat, cleaning the ears, and trimming the nails. Daily grooming should minimally include brushing the coat while checking all areas of the body. Nail trimming and ear cleaning may only be needed once every week, depending on the condition of the individual dog. Ideally, routine dental care is performed daily, and this may involve brushing the teeth or providing a dental chew toy. Cleaning the eyes depends on the presence and amount of discharge.

Daily Brushing

Daily brushing keeps a detector dog's coat clean. First brush with the lay of the coat, then loosen the undercoat with a fingertip massage, working against the lay of the coat. Brush with the lay of the coat to finish. On hard-coated dogs (dogs with short, tight coats with little fluffy undercoat), always brush with the lay of the coat. A variety of brushes or grooming tools can be used on different types of coats.

Rubber Curry Brush

These brushes or mitts are excellent for stimulating the skin and removing shedding coat. They are extremely gentle on the skin.

Rubber Slicker Brush

This is a good brush for daily brushing of short- and medium-coated dogs, as it is gentle on the skin and removes normal shedding coat.

Natural Bristle Brush

Natural bristle removes more of the shedding coat in short- and medium-coated dogs and spreads natural oils from the skin through the hair.

Hound Glove

A hound glove is a mitt with horsehair bristles attached to one side. It is excellent for tight, hard coats with little undercoat and for a quick "polish" before taking the dog onto the floor to work.

Other Grooming Tools

 The following tools are probably never appropriate for use on beagles or pointers and other smooth coated breeds, but are useful for grooming heavy coated breeds (Labrador retrievers, golden retrievers, springer spaniels).

Wire Slicker Brush

A wire slicker brush is appropriate for heavy-coated breeds (shepherd dogs, some Labrador retrievers) with substantial undercoat. Apply gentle pressure when using, as the tines can scratch or irritate the skin if applied too vigorously. Brushing gently against the lay of the coat will remove shedding coat more effectively than brushing with the lay of the coat with heavy pressure.

Undercoat Comb

An undercoat comb has long and short, offset, alternating tines. This tool is excellent for removing thick, shedding undercoat from heavy-coated breeds. In medium-coated breeds it can be used with caution to remove wadded undercoat. It can also be used to pick out matted coat in breeds with long, silky coats (golden retrievers, springer spaniels).

Mat Splitters (combs, rakes, blades)

These tools feature sharpened blades as part of the comb or rake configuration. They should be used only to remove matted hair from breeds with long, silky coats (golden retrievers, springer spaniels).

Conditioning the Coat

A wide array of commercial coat conditioners is available to polish or soften dogs' coats. One of the most effective ways to condition a dog's coat is by stroking the coat, with the grain, with the palms of your hands. Commercial conditioners that contain silicones are effective for repelling dust and stains and impart a high gloss to the coat, whereas those containing natural oils (emu, mink) are best for moisturizing dry skin and dry, brittle coats.

Trimming the Nails

Nails of detector dogs should be kept short for a well-kept appearance, but even more important, to prevent unnatural positioning of the feet and strain on the limbs. Long nails compromise the dog's surface traction (which is derived from surface contact with the pads), and can place unnecessary strain on the bones and ligaments of the paws.

Scissors and guillotine trimmers are available and the choice of which to use depends on handler preference. Motor-driven nail grinders are an excellent tool for keeping nails tightly groomed, but carry the disadvantage of poor acceptance by dogs not accustomed to them. Novice nail trimmers, or handlers with dogs whose nails are black can safely trim nails using a coarse file or cutting using the "deli slice" method (removing several thin slices of nail, rather than making one thicker cut). Ideally, the nail is trimmed to within 2 mm of the quick (fleshy nail bed). Handlers should aim for cutting the nails short, but not so short as to cut into the quick, which causes the dog pain, bleeding and apprehension about future nail trimming. Nails should be regularly trimmed, as the quick and the nail's blood supply lengthen when the nails are neglected, and only recede as the nails are gradually trimmed short. Handlers should be aware that they must also trim the dewclaws on those breeds with these nails intact.

Health Care: Grooming
Trimming the Nails

The following directions and illustrations were provided by Dr. Rose Borkowski:[1]

1. Note how the blood supply gets longer as the nail grows. Refer to **Figure 3-5-1**. If the nail is cut at ①, it will cause bleeding and the nail will still be long. Trim at ②, and the blood supply will begin to recede.

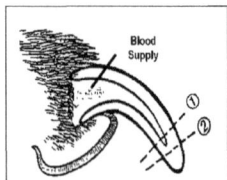

FIGURE 3-5-1: Blood Supply of a Dog's Nail

2. If a little is clipped or filed off a long nail every few days, the blood supply will continue to recede. As the nail get shorter, so does the blood supply. Refer to **Figure 3-5-2**.

FIGURE 3-5-2: Blood Supply Recedes as the Nail is Trimmed

3. Only when the nail is kept short can it be trimmed without bleeding. Refer to **Figure 3-5-3**. If the same cut had been made on the long nail illustrated in **Figure 3-5-1**, it would have caused bleeding.

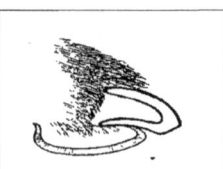

FIGURE 3-5-3: Properly Trimmed Nail

---

[1] Dr. Rose Borkowski. staff doctor at Tufts University. MA, specializing in wildlife exotics from 1994 to present; private practice in Boca Raton, 1994; attended University of Florida Veterinarian School, 1991.

Health Care: Grooming
Cleaning the Ears

Cleaning the Ears

Following are basic steps for cleaning a dog's ears:

1. Pull the earflap upward to expose the external opening of the ear canal.

2. Place several drops of ear cleaner solution in the dog's ear. Refer to *Figure 3-5-4*.

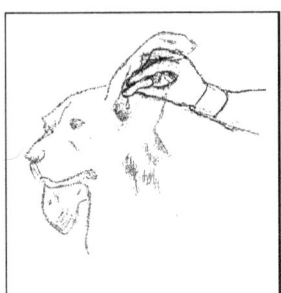

FIGURE 3-5-4: Cleaning Ear

3. Release the earflap.

4. Massage the base of the ear. Refer to *Figure 3-5-5*.

FIGURE 3-5-5: Massaging Ear

5. Repeat Steps 1–5 for the other ear.

6. Allow the dog to shake its head. Refer to *Figure 3-5-6*.

FIGURE 3-5-6: Allowing the Dog to Shake its Head

7. Pull the earflap upward, wipe the ear dry, and remove excess ear wax. Refer to **Figure 3-5-7**.

FIGURE 3-5-7: Wiping the Ear

Floppy-eared dogs are prone to yeast infections in their ears.

Cleaning the Eyes

A small amount of thin, clear discharge is normal for many dogs, especially for beagles with large, prominent eyes. If the discharge is yellow, green, or thick, it should be checked by a veterinarian. Beagles' eyes often discharge a thin fluid that creates a dark reddish-brown "tear" stain in the corners of the eye. This should be cleaned for both cosmetic and for health reasons. Any discharge that remains on the face is a potential breeding ground for bacteria. Eyes can be wiped with a soft, moist gauze sponge, or cleaned with a commercial tear stain remover if staining of the coat surrounding the eyes has occurred.

Cleaning the Teeth

Detector dogs should have their teeth cleaned a minimum of once a year, or more often, based on the veterinarian's recommendation. Rawhide chews function about as well as brushing to rid a dog's teeth of plaque (the precursor of the hard calculus deposits). The disadvantage of using rawhide is that it may cause gastrointestinal obstruction. Compressed rawhides are a good alternative, as these beak down more easily than rawhide strips. Brushes, finger brushes and special toothpastes are available for dogs. A moistened gauze pad will provide adequate, gentle abrasion to remove plaque from the teeth if used daily. NEVER use human toothpastes, as these contain foaming agents and other components that may cause digestive problems in dogs. Clean the outside area of the teeth, only. Saliva adequately clears the inner surfaces.

Health Care: Grooming
Expressing Anal Glands

Expressing Anal Glands

For information on expressing anal glands, refer to **Appendix G**.

Bathing

Determine how often the detector dog requires a bath, based on its skin needs. Keep in mind that excessive bathing is drying to the dog's skin and coat.

When bathing, select a mild shampoo made for dogs. For skin conditions, follow the veterinarian's advice about the type of shampoo to use and how to apply it. Flea and tick shampoos are not necessary for dogs receiving the monthly topical preventive (Frontline, Advantage, etc.), and, in fact, may be harmful to these dogs.

Important

Do not bathe the dog in cold or wet weather unless you can dry the dog in a warm place.

Secure the dog in the bathtub before bathing. **Do not leave the dog unattended.**

The basic bathing procedure is as follows:

1. Perform daily health inspection.
2. Brush coat and remove excess hair with fingertip massage.
3. Inspect for presence of external parasites (flea excrement, other insects, ticks).
4. Place a drop of mineral oil or triple antibiotic ointment in eyes.
5. Dispense shampoo into the hands and apply it to the dog's body.
6. Wet the coat with water and massage into a lather.
7. Let the shampoo stay on the dog for the time recommended by the shampoo manufacturer.
8. Rinse the coat thoroughly with warm water.
9. Massage the dog briskly with a dry towel.
10. Brush out excess hair.

For special bathing problems such as tar, sap, gum, or other sticky materials adhering to the coat, trim out as much of the substance as possible, and soak any residue in vegetable oil for several hours (or overnight for particularly resistant substances). Never use petroleum-based oil products on a dog's coat, as these can be ingested by the dog as it attempts to groom itself.

Health Care: Grooming
Bathing

4 Training

National Detector Dog Manual

Contents

Introduction **page 4-1-1**
  Selection **page 4-1-1**
  BCOT Requirements **page 4-1-2**
  Training Components **page 4-1-2**
Canine Officer Supervisor's Training (COST) **page 4-1-11**
  Training Objective **page 4-1-11**
  Estimated Time **page 4-1-11**
  Training Materials **page 4-1-11**
  Training Topics **page 4-1-12**
PPQ Officer Training **page 4-1-12**
  Training Objective **page 4-1-12**
  Estimated Time **page 4-1-12**
  Training Materials **page 4-1-12**
  Training Topics **page 4-1-12**

Introduction

The Basic Canine Officer Training (BCOT) course is held at the National Detector Dog Training Center (NDDTC) in Orlando, Florida, and is a requirement for all Canine Officers. The BCOT course is approximately 10 weeks. Students learn the basic skills necessary to perform their duties as Canine Officers. The course is structured to accommodate various pathways of AQI operations including passenger clearance, cargo control, and border control. The well-known Beagle Brigade is the force that works with passenger clearance. Larger breed dogs are used for the border and cargo pathways.

Selection

PPQ officers interested in becoming Canine Officers should request information about the agency's detector dog program and the availability of positions through their port directors. Also, PPQ officers can apply through merit promotion vacancy announcements or lateral transfer announcements. It is now a requirement that new Canine Officers commit to a period of time in the detector dog program.

Port Directors select Canine Officer candidates and notify the RCPC of the selections. The RCPC contacts the NDDTC to schedule training for Canine Officer candidates.

Training:
Introduction

Canine Officers are in the PPQ Officer, 436 series. Therefore, Canine Officer candidates must first meet New Officer Training (NOT) conditions of employment and complete all NOT and port requirements before attending BCOT.

BCOT Requirements

Weekly Evaluations

Each student's progress is evaluated weekly, documented, and reviewed with a trainer. This evaluation is an interactive process with the trainer's role being that of mentor and progress facilitator.

Students must achieve written test scores of at least 80 percent and successfully pass a validation test that demonstrates mastery of practical scent detector dog handling skills. After successfully completing BCOT, Canine Officers must complete annual validation testing.

Training Components

The BCOT course has many components, which are described below.

- **Pre-course Work**
- **Week 1**
- **Weeks 2 Through 5**
- **Weeks 6 Through 9**
- **Week 10**
- **Validation, Graduation, and Departure**

Pre-course Work

Thirty days prior to attending BCOT at the NDDTC, Canine Officer candidates receive an orientation packet with pre-course work. They must complete this work before arriving at the NDDTC.

Week 1

Instruction is held in a traditional classroom environment at the NDDTC. On the first day, students are oriented to the course content, performance expectations, and the NDDTC, as well as with policies covering standards of conduct and dismissal. A vehicle is assigned to the students while attending the course.

During the first week, exams are administered on the pre-course work after a brief review. The following topics are covered during the first week:

- **Canine Anatomy**
- **Canine Health**
- **Canine Behavior**
- **Safety**
- **Kennel Procedures**
- **Abuse**
- **Rapport Training**

Training:
Introduction

Canine Anatomy	Students learn the basic terminology necessary for discussing canine structures relevant to scent detection work (i.e., olfactory apparatus) and health (e.g., hips, spine, other skeletal landmarks, placement of internal organs). Mastering this terminology is necessary for precise communication about canine health, maintenance, and handling directions.
Canine Health	Instruction covers common canine diseases, parasites, injuries, and conditions. The focus is on preventive care (e.g., routine veterinary care, grooming, kennel hygiene, environmental considerations, maintaining healthy working weight) to avoid health problems that could affect a dog's performance. Students learn to recognize symptoms of health conditions that could have an impact on a dog's performance.
Canine Behavior	Students learn the basics of canine behavior, which is the foundation for all training. This training component covers fundamental concepts relating to canine behavior (e.g., behavioral tendencies, instincts, requirements, treatment, types of conditioning and learning, reward schedules, motivation) as they apply to handling a scent detection canine and troubleshooting performance-related difficulties. Time is devoted to sharpening students' abilities to observe and analyze canine behavior, emphasizing skills that will help them keep their detector dogs safe, healthy, and working proficiently.
Safety	Safety issues relate to both dog and handler in the training environment and beyond, when the detector dog team returns to its port. The primary topics of this component are: ◆ How to avoid accidents through heightened awareness of surroundings ◆ How Canine Officers can protect their detector dogs from harm while working around passengers, carts and luggage
Kennel Procedures	Instruction covers kennel procedures (e.g., one-way system, sanitation, exercising dogs, grooming schedules) primarily to keep Canine Officers and their detector dogs from unnecessary injury or stress while attending training at the NDDTC. Permanent kennel environments are unique to each port; however, the hygiene, health, and safety procedures used at the NDDTC are a standard by which students can measure the kennel facilities and procedures at their work locations.
Abuse	This section covers standards for the treatment of detector dogs and guidelines concerning what constitutes abuse or neglect of dogs.
Rapport Training	Upon successful completion of the classroom work and exams, students are assigned their detector dogs. Students begin their new relationships with their detector dogs by learning how to establish a rapport. Activities include hands-on experience such as grooming their dogs (e.g., bathing, cleaning ears, trimming nails).

Training:
Introduction

Weeks 2 Through 5 Instruction is held in a large, training area at the NDDTC. Suitcases, vehicles, and a mail conveyor belt are used to simulate work situations that will be encountered by students when they return to their work locations. The following topics are covered during these four weeks:

- **Motivation and Rewards**
- **Teamwork**
- **Lead Control**
- **Stimulus Control**
- **Voice**
- **Training Exercises**
- **Search Techniques**
- **Mail Facility Inspection Training**

Motivation and Rewards Students learn what motivates their dog and how to reward their dog.

1. Observation—Students learn to observe their dogs to determine what rewards are most likely to elicit or enhance their strong desire, persistence, and enthusiasm to work.

2. Canine preferences—Although the standard reward is food, some dogs perform most effectively when food is combined with other types of rewards or reinforcers (e.g., towel, ball, etc.). Students learn to distinguish between primary and secondary reinforcers and use this information to create a meaningful reward system for their detector dogs.

    A. Primary—Reinforcers that are intrinsic, based on physical needs (e.g., food, sex, touch) or because of genetic predispositions of the breed (prey drive, visual stimuli).

    B. Secondary—Reinforcers that are related to conditioning and have a context in which an association has been learned through experience (verbal praise, smiles, toys, etc.).

3. Timing—Students learn the importance of timing in reward delivery.

Teamwork

Teamwork is central to the relationship between dog and handler and is an ongoing experience throughout BCOT. Students must interface with their dogs and their classmates to accomplish daily tasks. This includes delegating tasks, accepting responsibility, and enabling each member of the team to reach his or her potential in performing assigned tasks. The teamwork skills students acquire are applicable to the work environment where a team approach enhances work efficiency, such as working with other inspection personnel in the port. Teamwork is so pivotal to creating successful detector dog teams that it is formally addressed, then applied throughout the course. Students learn to:

1. Delegate canine tasks—In the working detector dog team, the dog is delegated the task of locating target items by its sense of smell. This requires the handler (student) to relinquish control of that task and allow the dog to do its job.

2. Delegate handler tasks—In the working detector dog team, the handler is delegated the tasks of protecting the dog, observing the dog for indications that it has detected odor, and facilitating the dog's efforts to locate an odor source. This activity requires the student to accept responsibility for those tasks.

3. Lead a team—In any team effort, the leader assesses team member potentials, coordinates and delegates tasks, and is ultimately responsible for the safety and effective functioning of the team.

4. Care for kennel as a team—The students work in teams with one another and with animal care technicians, applying their teamwork skills daily when maintaining the training area and performing kennel care responsibilities.

Lead Control

The lead is an important communication device between the handler and the dog. As such, proper deployment of a lead is critical to developing a successful working relationship.

1. Basics—Students learn what types of leads are most appropriate for scent detection work and understand the differences in the types of leads.

2. Techniques—Students learn the proper use of a lead, how to hold a lead and manipulate it to achieve control and communication, how to keep the appropriate tension on the lead, and how to use the lead as a guide to ensure maximum coverage of articles being inspected.

Training:
Introduction

Stimulus Control
: Students learn how to use certain stimuli to control their dog's behavior. Students learn:

  1. Specific commands to elicit behaviors from their dogs.
  2. Behaviors of discrimination and generalization and how to balance these two behaviors to achieve optimal detection efficiency.
  3. Incidental stimuli that may exert control over a dog's responses (e.g., environment, uniformed handler, and inspection center venue).

Voice
: Quality, pitch, and tone of voice convey meaningful information to dogs. Students learn the importance of consistency in their use of words, and to identify the effects of different voice qualities. Aspects of voice control include the following:

  1. Commands—Students learn a specific set of commands or signal words. Since the dogs have already learned to associate these words with performing certain activities, consistent use of these commands will facilitate the transition of the dogs to their new handlers.
  2. Timing—To become established as a meaningful cue that the dog associates with a behavior, a verbal cue must be delivered at the appropriate time and within a narrow time window. Students learn the logic of contingency as it relates to canine learning.
  3. Obedience—The willingness of dogs to perform certain tasks can be enhanced by adjusting the tone of voice. Students learn how to apply this skill.
  4. Inflection—Students learn how to use and emphasize certain sounds to maximize the attention of their dogs.

Training Exercises
: Students gain experience in how to create valid training exercises by assisting in setting up exercises for the class. These skills and guidelines will apply when students continue training their detector dogs at their work locations. To conduct valid, useful training, students are made aware of the following issues:

  1. Cross contamination—How to preserve the integrity of the dog's response to specific target odors.
  2. Odors (target/non-target)—Which odors constitute target, which are non-target, and how to introduce the dog to new odors in training.
  3. Containers—How to safeguard and protect the integrity of training aids.
  4. Documentation—Record keeping, using worksheets to track the training progress of the dog.

Search Techniques	Once students have mastered basic presentation skills, they are introduced to techniques that will enhance their work efficiency with their detector dogs. These techniques include the following:

1. Search Patterns—How to move a dog and move with a dog around typical obstacles and luggage to maximize coverage of the search.

2. Breathing Bags—How to assist a dog in detecting odors within luggage by pushing air out of the bags at the appropriate time.

3. Tap Backs—How to provide a dog a second chance to examine a piece of luggage without interrupting the flow of the search.

4. Pinpointing—How to induce a dog to be specific when indicating on an odor source either by touching with its nose or with its paw.

Mail Facility Inspection Training	This training component introduces students to techniques for searching parcel post and packages on a conveyor belt. Even if their work locations have no current plans to use a detector dog team at a mail facility, students will benefit from this training for off-task exercises.

1. Acclimate dog to belt—Dogs must become accustomed to the noise and movement of a conveyor belt and must be comfortable working on the moving surface.

2. Use active response—For increased safety and efficiency, dogs are encouraged to actively indicate by touching or scratching the parcel.

3. Set up exercises—Students learn how to create realistic training scenarios peculiar to the circumstances encountered in mail facilities.

Weeks 6 Through 9	The instruction presented during weeks 6 through 9 is specialized for a detector dog team's application pathway. The training occurs at an international airport, an international border, or an international cargo receiving area that closely simulates the environment in which the detector dog teams will eventually work. During these four instructional weeks, students continue to perfect their skills by applying them to the appropriate environment. The following skills are covered during these four weeks:

- **Pathway Application**
- **Troubleshooting**
- **Travel**
- **Media**
- **Monthly Reports**

Training:
Introduction

Pathway Application

This training component might be considered on-the-job training. It occurs in two phases. The first phase is conducted at the NDDTC where procedures peculiar to each scenario are explained and discussed. Role-playing and preparation for public contact expected when the detector dog teams are deployed in the field are reviewed. Paperwork particular to each pathway is reviewed.

The second phase occurs at an international airport, an international border, or an international cargo receiving area where practical handling skills are applied in real-life scenarios.

1. Passenger Clearance
    A. Students learn how to deal effectively and efficiently with passengers in a Federal Inspection Service facility while maintaining the safety of their detector dogs and themselves. Students learn how to do the following:
        i. Work safely—Safety is addressed for both dogs and handlers, and common hazards are described.
        ii. Search carts, high/low—Students learn how to present areas during searches and detector dogs learn how to indicate finds of target odors in and around baggage carts.
        iii. Interview passengers—Students learn how to safeguard their detector dogs while interviewing passengers, how to ask specific questions to facilitate the interview process, how to verify the contents of baggage, and how to deal diplomatically with difficult passengers.
        iv. Recognize residual odors—Target odors in baggage that recently contained agricultural items will elicit a response from detector dogs. Students are instructed in recognizing this type of response.
        v. Vary reward schedule—Students learn about the effectiveness of variable reward schedules.
        vi. Process paperwork—Students learn how to use PPQ Form 277, Baggage Information Data Card, and how to process Customs Declarations (Customs Form 6059B).
    B. Students learn how to present their detector dogs to maximize the positive impression of the agency in public relations functions and for media events.

2. Border

   A. Vehicle or conveyance—Students learn how to search areas of vehicles (e.g., buses, cars, vans, trailers) that are likely to contain contraband.

   B. Luggage—Students learn how to deploy their detector dogs on luggage and search these after their indications.

   C. Process paperwork—Students learn how to use PPQ Form 277, Baggage Information Data card and how to process Customs Declarations (Customs Form 6059B).

   D. Safety and health issues—Students learn about environmental concerns (e.g., temperature, noxious fumes) and their relevance to canine health and safety.

3. Cargo

   A. Warehouses—Students learn about search patterns and techniques that apply when large, enclosed areas with boxes, bulk stock, and pallets are to be searched.

   B. Containers—Students learn about special problems that may be encountered with their detector dogs when searching shipping containers containing agricultural items in various conditions (such as in bulk or refrigerated, from ships or trucks).

   C. Manifests—Students are familiarized with documentation involved in processing cargo and learn how to screen cargo shipments to maximize the effective use of their detector dogs.

   D. Safety and health issues—Students learn about environmental concerns (e.g., temperature, noxious fumes, stability of stacked boxes and shipping materials) and their relevance to canine health and safety.

Troubleshooting Students are initially guided through the process of applying the principles of conditioning to correct common work deficiencies or behavioral problems with their detector dogs. They are expected to quickly acquire and apply this skill independently. Following are some of the procedures students are required to learn before returning to their work locations:

1. Problem analysis—Students must examine and identify the elements of behavioral scenarios and scent work challenges. Commonly encountered problems such as regression, distractions, and performance issues are discussed.

2. Corrective action plan—After analyzing a problem, the student will be able to apply principles of conditioning to work towards a solution.

Training:
Introduction

        3. Documentation—Students learn how to document problems and corrective measures, and how to identify problems using monthly documentation.

        4. Environmental distractions—Students learn to deal with conditions present in the work place, such as physical obstacles, noxious odors, noisy machinery, other animals, air currents, temperature extremes, and unruly passengers.

Travel

Students learn to prepare their detector dogs to travel and return to their work locations. Instructions include necessary paperwork, health examinations, records, flight restrictions, letters of acclimation, equipment needed, recommended temperatures, water, bedding, and types of crates.

Media

Students are given general guidelines for presenting information about the Detector Dog Program, including the Beagle Brigade, and USDA programs to the public. Topics covered include presentation approaches for different age groups (e.g., demonstrations, lectures, and interactive activities), common pitfalls, and access to handouts.

Monthly Reports

Students are taught to enter and manage data to produce monthly reports that are submitted to the RCPC (and NDDTC staff for the first six months).

Week 10

The final week is an assessment of all the skills mastered to ensure the detector dog teams are equipped to meet the mission needs of their work locations. A mock validation exercise is conducted in the beginning of the week followed by a day to correct any deficiencies noted. A final validation exercise is conducted towards the end of the week. Students who pass the validation exercise receive training completion certificates at a graduation ceremony held at the end of the week.

Validation, Graduation, and Departure

Students will participate in the following activities:

1. Mock validation—A validation exercise is set up to familiarize students with the procedure for the practical exam and allow them an opportunity to practice.

2. Proficiency training—Deficiencies noted in the mock validation are addressed and corrected.

3. Validation—The practical exam, in which detector dog teams demonstrate mastery of skills (i.e., handling, safety, behavioral assessment, etc.) and teamwork in a realistic scenario. Students may be re-tested one time (at the discretion of their instructor) if they do not pass the validation test.

4. Graduation (generally held in the location of Validation).

5. Transition to work locations—Students learn how to transfer the skills their detector dogs have acquired to their work locations. Topics covered include: overcoming relocation stress, acclimating to new work environments, handling abuse or interferences, training to overcome regression, increasing endurance, conditioning, documenting team activities, acquiring and maintaining equipment, and establishing local suppliers.

6. The NDDTC staff provides additional support for 6 months after the detector dog team completes BCOT. This support is provided in coordination with the RCPC, who is responsible for installation assessments and follow-up contacts.

Canine Officer Supervisor's Training (COST)

COST is training and orientation developed for supervisors of detector dog teams covering all aspects of managing detector dog teams. The training is scheduled and conducted by a Regional Canine Program Coordinator (RCPC) at the work location.

Training Objective

The objective is to provide standardized training and orientation to supervisors of detector dog teams.

Estimated Time

The duration of training is from 12-16 hours depending on class size and whether training will also be presented to PPQ officers. See ***PPQ Officer Training***.

Training Materials

The following materials are used in training:

- Instructor Guide for Canine Officer Supervisor's Training
- Set of overheads
- Set of handouts
- *National Detector Dog Manual*

Training Topics

The following topics are covered in the training:

- Roles and responsibilities that support the Detector Dog Program
- Work expectations and team development
- Utilization of detector dog teams
- Equipment and supplies
- Veterinarian requirements
- Proficiency training
- Outreach information and responding to the media
- Documentation and reporting

PPQ Officer Training

Training and orientation developed for PPQ Officers who are coworkers of detector dog teams. The training is scheduled and conducted by an RCPC at the work location.

Training Objective

Provide standardized training and orientation to coworkers of detector dog teams.

Estimated Time

The duration of training is from 1-2 hours depending on class size and whether training will also be presented to supervisors. See *Canine Officer Supervisor's Training (COST)*.

Training Materials

- Instructor Guide for PPQ Officer Training
- Set of overheads
- Set of handouts
- *National Detector Dog Manual*

Training Topics

The following topics are covered in the training:

- Roles and responsibilities that support the Detector Dog Program
- Utilizing detector dog teams

Glossary

Air-scenting	Behavior exhibited when a dog uses air currents to lead itself to the source of the odor. For example, a dog lifts its head, actively smelling, and crosses the terminal to a person eating an apple.
Alert response	A trained physical response to a stimulus. For example, when a detector dog sits at a target bag.
ALT	Alanine transferase.
AWA	Animal Welfare Act.
Biological break	Time to take a detector dog to an area where the dog can urinate and/or defecate. While a detector dog is housed in a secondary residence (crate or wire kennel), allow a biological break a minimum of once every 2 hours. Plan for at least a 15 minute break.
Blank container	A container (suitcase, box, etc.) without contraband.
Blank exercise	A training problem containing NO target odors. Blank exercises can help identify whether or not a detector dog has a false response problem.
BUN	Blood urea nitrogen.
CBC	Complete blood count.
Contraband	Items of agricultural interest that are prohibited.
Correct response or positive response	An action by a detector dog correctly indicating an agricultural item.
Cuing dog	A verbal or physical action by a Canine Officer causing the detector dog to respond to an odor.
Ectoparasite	An external parasite that lives in or on the skin of the host.
Endoparasite	An internal parasite that lives within the body of the host.

Glossary

Exercise	Operationally, exercise means taking a detector dog out of either its secondary or primary residence for a period of no less than 5 minutes to either walk on a lead or a leash or to move about in a fenced-in, secure area. Never leave a detector dog unattended while exercising.
Extinction exercise	A training problem containing notable non-target odors but **no** target odors. Extinction exercises help resolve false response problems.
False response	An indication by a dog that it has detected a target item, when in fact, no target item or odor exists.
Food guarding	Behavior where a dog covers, envelops, or conceals its food and/or bowl when approached. Guarding can lead to aggression if challenged by a person or another dog.
Handbaggage	Luggage that is hand-carried on the plane, not checked as pit baggage.
Handler error	An action of a Canine Officer that causes the dog to make a mistake (false response, incorrect search patterns, pulling the dog off a target odor).
Hasty muzzle	A restraining appliance that fits over the dog's snout preventing biting used in emergency situations when no muzzle is available. Virtually anything can be used to make a hasty muzzle (belt, electrical cord, or a leash). Directions for applying a hasty muzzle are behind the green tab under ***Physical Restraint***.
Hyperthermia	Unusually high body temperature (overheating), as opposed to hypothermia.
Intermediate host	The animal or insect used by a parasite to develop its life cycle.
K/MLV	Killed or Modified Live Virus.
Mixed odor	Target and nontarget odors placed together in a container.
Mixed target odor	Variety of target odors placed together in a container.
NDDPM	National Detector Dog Program Manager.
NDDTC	National Detector Dog Training Center. See ***Appendix A***, ***APHIS Contacts***, for employee names, address, and phone number of the Center.
Nontarget odor	An odor that indicates no significant risk of pest or animal disease, such as leather or perfume; or an agricultural item, such as bread or fish. One that a dog has been trained not to respond to.

Odor generalization	Behavior where a dog indicates an odor to which it has not been trained to respond. For example, a dog has been trained to respond to pork and grapefruit. Then the dog responds to all meat and all citrus fruits. Also, a dog might generalize on nontarget items, such as fish, bread, or candy. Generalizing can be negative or positive.
OFA	Orthopedic Foundation for Animals, Inc.
Parasite	A living organism which, for the purpose of obtaining food, lives on or in a creature of a different species and causes harm or disease.
Pinpoint	Behavior where a dog goes directly to the source of the odor and indicates exactly where the item is. The dog usually will use its nose; some will use their foot.
Primary residence	The primary enclosure for the detector dog (boarding kennel). Must have a minimum amount of floor space as required by the AWA. The minimum amount of floor space required is the following:

$$\frac{\textit{Measurement (in inches) from the tip of the dog's nose to the base of its tail} + 6\ \textit{inches}}{144} = \textit{Required Square Feet of Floor Space}$$

The interior height must be at least 6" higher than the dog's head when the dog is in a normal standing position.

For example, a beagle 25" long (tip of nose to base of tail) and 17" tall (top of the dog's head when standing) would require a primary residence of 6.67 square feet of floor space with a height of 2 feet. The calculation is $(25" + 6")^2 / 144 = 6.67$; $17" + 6" = 23/24"$.

Primary reward	A reward to which a dog must have to survive (food, water, etc.) Food is the primary reward for PPQ's detector dog activities.
Proficiency rating	A measurement of the accuracy of a detector dog. In a training scenario it is obtained by dividing the number of positive responses by the number of total trials. In a working scenario it is obtained by dividing the number of positive responses by the total number of responses.
RCPC	Regional Canine Program Coordinator. See ***Appendix A***, ***APHIS Contacts***, for names, addresses, and telephone and fax numbers of RCPCs. See ***Appendix B***, ***Personnel***, for the roles and responsibilities of RCPCs.
Recovery time	The period a Canine Officer needs to rest the dog's nose after working a flight. The dog's nose becomes saturated with odor and is no longer able to detect target items. The recovery time differs from one detector dog to another and from one situation to another.

Redirected aggression	Hostile behavior displayed when a dog blocked from attacking another dog redirects its aggression towards a reachable target (bites a nearby person or handler).
Residual odor	Odor of a target item lingering in a bag after the item is no longer in the bag.
RPM	Regional program manager. See **Appendix A, APHIS Contacts**, for names, addresses, and telephone and fax numbers of the RPMs who manage detector dog activities in their region.
Saturation point	The point at which the detector dog can no longer detect additional sensory input.
Secondary residence	Secondary enclosure for the detector dog (sky crate, wire kennel) that is used for transport from the primary residence to the work site. The secondary residence also acts as the storage container for the dog while remaining at the work site. The container must be constructed of strong enough material to contain the dog securely and comfortably and withstand the normal rigors of transportation. The dog should not be able to put any part of its body outside the enclosure in a way that could result in injury to itself. The container should allow enough room for the dog to comfortably stand up, turn around, and lie down.
Secondary reward	A reward that is less significant than the primary reward. Examples are praise and petting.
Secure latch	Any added measure to secure a dog in its primary residence and to prevent it from accidentally or intentionally opening the kennel gate (for example, a clip).
Speed trials	A technique used to expedite the dog's final response to 1–2 seconds. Speed trials correct the problem of slow sits.
Target odor	An odor from a prohibited agricultural item that a detector dog has been trained to indicate.
Territorial behavior	Behavior displayed by a dog that protects or guards its particular kennel space if challenged by a person or another dog.
Training exercise	An entire training problem containing blank bags, nontarget bags, and target bags. An exercise should take place in a controlled environment.
Trial	The number of times a detector dog team inspects a target container in an exercise.
Validation testing	Testing administered to establish credibility, to assess the overall proficiency of the detector dog team, by identifying a team's strengths and weaknesses.

Appendix A

APHIS Contacts

Introduction

Use this appendix to locate and contact PPQ personnel who participate in detector dog activities. Note that the names of Canine Officers have been omitted to minimize updating this appendix.

Plant Protection and Quarantine

Port Operations

Donna L. West, National Detector Dog Program Manager (NDDPM)
Department of Homeland Security (DHS)
Customs and Border Protection (CBP)
Office of Field Operations – Agriculture Inspection (AI)
1300 Pennsylvania Avenue, NW
Rm. 5.4C-62
Washington, DC 20229
Ph. 202-927-0218
Cell Phone: 301-252-9191
Fx. 202-927-0116

National Detector Dog Training Center (NDDTC)

Jay Weisz, Director
Francisco (Frank) Ramos, Administrative Support Assistant
Lisa Beckett, Training Specialist
Monica Errico, Training Specialist
John Leuth, Training Specialist
Craig Schultz, Training Specialist
Michael (Mike) Smith, Training Specialist
Daniel (Dan) Talbert, Training Specialist
Carol Butler, Training Technician
Jennifer (Jenni) Hutto, Animal Caretaker
Alexis Soper, Animal Caretaker
Irene Taylor, Animal Caretaker

10806 Palmbay Drive
Orlando, FL 32824
Ph. 407-816-1221
Fx. 407-816-0690

Appendix A: APHIS Contacts
Plant Protection and Quarantine

Eastern Region

Eugene Jowyk, RPM
DHS, CBP, AQI
920 Main Campus Drive, Ste 200
Raleigh, NC 27606
Ph. 919-716-5724
Fx. 919-716-5656

Alison Pae, RCPC
USDA, APHIS, PPQ
Hartsfield International Airport
12700 Spine Road, Concourse E
Atlanta, GA 30320
Ph. 404-564-2313
Fx. 404-564-2314

Albert Roche, RCPC
USDA, APHIS, PPQ
IBM Building
654 Munoz Rivera Avenue, Suite 700
Hato Rey, PR 00918
Ph. 787-771-3611
Fx. 787-771-3613

Western Region

Bob Parker, RPM
DHS, CBP, AQI
2150 Centre Ave, Building B - 3E10
Fort Collins, CO 80526-8117
Ph. 970-494-7562
Fx. 970-494-7501
Cellphone 970-988-3506

Grace Nagano, RCPC
USDA, APHIS, PPQ
Honolulu International Airport
300 Rodgers Boulevard, #57
Honolulu, HI 96819
Ph. 808-861-8490
Fx. 808-861-8501

Appendix A: APHIS Contacts
Plant Protection and Quarantine

Work Locations—Eastern Region

Atlanta, GA
USDA, APHIS, PPQ
P. O. Box 45408
Atlanta, GA 30320
Ph. 404-564-2290, ext 1111
Fx. 404-564-2285

Boston, MA
USDA, APHIS, PPQ
Logan International Airport Terminal E
East Boston, MA 02128
Ph. 617-568-1481
Fx. 617-561-5739

Buffalo, NY
USDA, APHIS, PPQ
783 Busti Avenue, First Floor
Buffalo, NY 14213-2405
Ph. 716-881-5755
Fx. 716-551-3976

Charlotte, NC
USDA, APHIS, PPQ
1901-A Cross Beam Drive
Charlotte, NC 28217
Ph. 704-359-4772
Fx. 704-359-4766
Bp. 888-698-9407

Chicago, IL
USDA, APHIS, PPQ
O' Hare Airport Station
P.O. Box 66192
Chicago, IL 60666
Ph. 773-894-2920
Fx. 773-894-2927

Detroit, MI
USDA, APHIS, PPQ
Metro Airport
Airport Operations
International Terminal, Room 228
Detroit, MI 48242
Ph. 734-942-7024
Fx. 734-942-7409

Dulles, VA
USDA, APHIS, PPQ
Dulles International Airport
P.O. Box 17134
Washington, DC 20041
Ph. 703-661-8348
Fx. 703-661-8165

Appendix A: APHIS Contacts
Plant Protection and Quarantine

Erlanger, KY   USDA, APHIS, PPQ
Greater Cincinnati/N. Kentucky Airport
International Terminal, Concourse B
P.O. Box 18402
Erlanger, KY 41018
Ph. 859-767-7070
Fx. 859-767-7074

Jamaica, NY   John F. Kennedy International Airport
USDA, APHIS, PPQ
JFKIA IAT Room 2317
Jamaica, NY 11430
Ph. 718-244-2175
Fx. 718-553-0092
    718-553-1796 Kennel

Memphis, TN   USDA
8001 Centerview Parkway
Suite 216
Cordova, TN 38018
Ph. 901-797-7750
Fx. 901-544-0375

Miami, FL   USDA, APHIS, PPQ
13631 Old Cutler Road
Miami, FL 33158
Ph. 305-232-9549 Kennel
Ph. 305-869-3162 Airport
Fx. 305-251-8944

Minneapolis, MN   USDA, APHIS, PPQ
Minneapolis/Saint Paul Airport
P. O. Box 11690
Saint Paul, MN 55450
Ph. 612-725-0078
Fx. 612-727-2442

Newark, NJ   USDA, APHIS, PPQ
Newark International Airport
Terminal B, International Arrivals Area
Newark, NJ 07114
Ph. 973-645-6194
Fx. 973-645-6389

Orlando, FL   USDA, APHIS, PPQ
9317 Tradeport Drive
Orlando, FL 32827
Ph. 407-648-6856
Fx. 407-648-6859

Appendix A: APHIS Contacts
Plant Protection and Quarantine

Philadelphia, PA	USDA, APHIS, PPQ Philadelphia International Airport Richardson Dilworth, Terminal A Philadelphia, PA 19153 Ph. 215-596-4784 Fx. 215-596-0698
Ponce, PR	USDA, APHIS, PPQ Ponce Work Unit Terminal Building P. O. Box 45 Mercedita, PR 00715 Ph. 787-841-3225 Fx. 787-841-3195
Port Huron, MI	USDA, APHIS, PPQ 2321 Pine Grove #124 Port Huron, MI 48060-1306 Ph. 810-985-6126 Fx. 810-985-5542
San Juan, PR	USDA, APHIS, PPQ LMM International Airport P.O. Box 37521 Airport Station San Juan, PR 00937 Ph. 787-253-4506 Fx. 787-253-4646
Tampa, FL	USDA, APHIS, PPQ Tampa Work Station 4951-B East Alamo Drive Suite 220 Tampa, FL 33605 Ph. 813-228-2121 or 2122 Fx. 813-228-2441

Appendix A: APHIS Contacts
Plant Protection and Quarantine

Work Locations—Western Region

Blaine, WA
USDA, APHIS, PPQ
P. O. Box 1930
Blaine, WA 98231-1930
Ph. 360-332-8891
Fx. 360-332-7830

Brownsville, TX
USDA, APHIS, PPQ
Veterans Int'l Bridge
3300 Expressway 77/83
Room A-151
Brownsville, TX 78521
Ph. 956-983-5800
Fx. 956-983-5830

Calexico, CA
USDA, APHIS, PPQ
P. O. Box 2940
Calexico, CA 92231
Ph. 760-768-2540
Fx. 760-768-2546

Dallas, TX
USDA, APHIS, PPQ
P.O. Box 610063
DFW Airport, TX 75261
Ph. 972-574-2117
    972-574-9605
Fx. 972-574-5258

Denver, CO
USDA, APHIS, PPQ
3950 North Lewiston, Suite 330
Aurora, CO 80011
Ph. 303-371-3355
Fx. 303 371-4774

El Paso, TX
USDA, APHIS, PPQ
Cordova Border Station
3600 E. Paisano, Rm. 154-A
El Paso, TX 79905
Ph. 915-872-4720
Fx. 915-534-6653

Honolulu, HI
USDA, APHIS, PPQ
300 Rodgers Blvd.
Honolulu International Airport
Terminal Box 57
Honolulu, HI 96819
Ph. 808-861-8490
Fx. 808-861-8501

Houston, TX USDA, APHIS, PPQ
3004 Mecom Road
Houston, TX 77032
Ph. 281-233-3670
Fx. 281-233-3678

Laredo, TX USDA, APHIS, PPQ
New Border Station, Rm. 505
Lincoln-Juarez Bridge Building 5
Laredo, TX 78042
Ph. 956-726-2225
Fx. 956-726-2322

Long Beach, CA USDA, APHIS, PPQ
11 Golden Shore, Suite 460
Long Beach, CA 90802
Ph. 562-980-4222
Fx. 562-980-4208

Los Angeles, CA USDA, APHIS, PPQ
11840 LaCienega Blvd.
Hawthorne, CA 90250
Ph. 310-215-2431
Fx. 310-215-1379

Maui, HI USDA, APHIS, PPQ
1 Kahului Airport Road, Unit 11
Kahului, HI 96732
Ph. 808-877-8757
Fx. 808-877-9086

Nogales, AZ USDA, APHIS, PPQ
9 North Grand Avenue, Room 2214
Nogales, AZ 85621
Ph. 520-287-4783
Fx. 520-287-6941

Oakland, CA USDA, APHIS, PPQ
1301 Clay Street
Federal Building
Suite 160 North
Oakland, Ca 94612
Ph./Fx. 510-273-7254

Appendix A: APHIS Contacts
Legislative and Public Affairs

Pharr, TX USDA, APHIS, PPQ
Pharr International Bridge
9901 South Cage Blvd., Ste. A
Pharr, TX 78577
Ph. 956-283-2160 Pharr
Ph. 956-843-2552 Hidalgo
Fx. 956-783-5387

Phoenix, AZ USDA, APHIS, PPQ
3658 East Chipman Rd.
Phoenix, AZ 85040
Ph. 602-414-4740
Fx. 602-414-4775

San Diego, CA USDA, APHIS, PPQ
P. O. Box 434419
San Diego, CA 92143-4419
Ph. 619-662-7333
Fx. 619-662-7335

San Francisco, CA USDA, APHIS, PPQ
P.O. Box 250009
San Francisco, CA 94125
Ph. 650-876-2840
Fx. 650-876-0915

Seattle, WA USDA, APHIS, PPQ
SEATAC International Airport
Room 106 South Satellite
Seattle, WA 98158
Ph. 206-244-4244
Fx. 206-764-3825

Legislative and Public Affairs

Sue A. Challis, Public Affairs Specialist
Customs and Border Protection
Department of Homeland Security
Ph. 202-927-1547
Fx. 202-927-1393

Appendix B

Personnel

Contents

Introduction **page-B-1-1**
Canine Officers **page-B-1-2**
  Managing Detector Dogs **page-B-1-2**
  Training Detector Dogs **page-B-1-3**
  Maintaining Public Awareness and Communication **page-B-1-3**
  Monitoring Reports and Documentation **page-B-1-3**
Co-workers (Nonhandlers) **page-B-1-4**
Local Managers (Supervisors, Port Directors, and SPHDs) **page-B-1-4**
  Requesting New Detector Dog Teams **page-B-1-4**
  Managing Detector Dog Teams **page-B-1-5**
  Maintaining Public Awareness and Communication **page-B-1-5**
  Utilizing Detector Dogs **page-B-1-5**
  Monitoring Reports and Documentation **page-B-1-6**
Regional Canine Program Coordinators (RCPCs) **page-B-1-7**
Regional Program Managers (RPMs) **page-B-1-8**
National Detector Dog Instructors **page-B-1-9**
Animal Care Technicians at the NDDTC **page-B-1-10**
Professional Development Center (PDC) **page-B-1-10**
National Detector Dog Training Center (NDDTC) **page-B-1-11**
National Detector Dog Program Manager (NDDPM) **page-B-1-11**
Ensures that effective communication is a DDP norm **page-B-1-11**

Introduction

Use this appendix to identify the main roles and responsibilities that support detector dog activities in PPQ. The roles and responsibilities listed for the positions described are not all inclusive of the tasks performed by those who hold the positions.

PPQ's detector dog activities are managed within the regional structure by the RCPCs. At their assigned work location, the detector dog teams may be directed by a supervisor, manager, or port director. Supervisory and administrative support are provided through normal PPQ channels.

Appendix B: Personnel
Canine Officers

Those who support detector dog activities in PPQ hold the following positions:

- Canine Officers
- Co-workers (nonhandlers)
- Local managers (supervisors, port directors)
- RCPCs
- Regional program managers
- National Detector Dog Instructors
- Animal Care Technicians at NDDTC
- PDC
- NDDPM

Canine Officers

Following are the major responsibilities of Canine Officers organized by these categories:

- **Managing Detector Dogs**
- **Training Detector Dogs**
- **Maintaining Public Awareness and Communication**
- **Monitoring Reports and Documentation**

Important

In addition to other GS-436 performance elements, the following are examples of detector dog performance elements related to detector dog activities which can be used in part or modified for work unit utilization.

Some performance elements may combine various aspects of maintaining a detector dog or may be separated into more specific elements.

The port directors are responsible for delegating and implementing performance standards based on Department, Agency, and contractual requirements. Contact an employee relations specialist and/or RCPC for assistance in updating and/or revising performance standards for Canine Officers.

Managing Detector Dogs

Quickly detect any deviation in a dog's behavior and is able to correlate it with the dog's productivity. Is sensitive to a dog's health and is able to detect signs of illness in order to get immediate medical attention, if necessary. Maintain good animal welfare standards including the proper health program, environment, diet, and housing.

Utilize the detector dogs in innovative ways and work for maximum periods on flights where a dog is most useful in detecting quarantine material interceptions (QMIs). Review flights with RCPCs and coordinate with supervisors to determine which flights will be worked. Is diligent in following up on positive and negative responses. Keep dogs on leashes or in kennels at all times unless in secure, fenced-in areas.

Display support for improved customer service and finds ways to improve the detector dog activities.

- Is able to work independently with little or no direct supervision.

Training Detector Dogs

Maintain the motivation of their dog and conduct appropriate training. Procure, use, and maintain training aids and supplies for the ongoing training of detector dogs in order to increase their proficiency. Set up an area with the proper environment to conduct training and follow standard operating procedures at all times. Provide technical assistance to peer Canine Officers in order to overcome problems that may exist in the proficiency and deployment of their detector dogs. Maintain credibility of the detector dog team by successfully passing validation test when given.

Maintaining Public Awareness and Communication

Communicate to schools, public groups, and/or other parties requesting information on the detector dog programs. Structure each presentation to meet the needs of the benefiting group.

Actively develop a network of contacts that results in demonstrations that highlight PPQ's mission and detector dog activities. Establish and maintain effective relationships with print and broadcast media representatives.

Maintain adequate stock of outreach information (bookmarks, coloring books, etc.) that serves to expand understanding and awareness. Order outreach information through the RCPC.

When queried by passengers, promote public awareness of PPQ's mission.

Monitoring Reports and Documentation

Keep daily records of inspectional activities, control measures, applied regulatory procedures, training records, cost effectiveness, detector dog behavior, detector dog health, and detector dog proficiency.

Complete monthly narrative and statistical reports on detector dog activities. Prepare and submit an electronic version **by the 10th of the following month accurate and complete** monthly narratives as required in all phases of work with little or no direct supervision.

Maintain an updated copy of health records for each dog.

(Refer to ***Appendix H***, ***Reporting and Documentation***, for samples of the records maintained by Canine Officers.)

Co-workers (Nonhandlers)

Coworkers:

- Understand and cooperate with local operational procedures established for detector dog activities.

- Periodically assist a Canine Officer in baggage inspection to which a detector dog responds on the spot. This assistance may be incorporated in a rotating position.

- Help new Canine Officers with on-the-job secondary inspection.

- Realize that new detector dog teams and experienced teams with a 2-week or more period of inactivity will initially send to secondary inspection many responses that may not yield agricultural seizures. Improvement of reliability is indicated with regularity of detector dog work.

- Hold declarations for new Canine Officers to review, if possible.

Local Managers (Supervisors, Port Directors, and SPHDs)

The major responsibilities of local managers related to detector dog activities are organized by these categories:

- **Requesting New Detector Dog Teams**
- **Managing Detector Dog Teams**
- **Maintaining Public Awareness and Communication**
- **Handling Media Events**
- **Utilizing Detector Dogs**

Requesting New Detector Dog Teams

Local managers:

- Request that feasibility studies to be conducted by an RCPC.
- Select new Canine Officers, with input from RCPCs.

Managing Detector Dog Teams

Local managers:

- Support the detector dog program.
- Must ensure that all kennel requirements are met in selecting a kennel. Choose a kennel, along with Canine Officers and the RCPC.
- Must ensure that kenneling and sanitization standards are being met. Visit the kennel at least twice a year.
- Help to select a veterinarian, along with Canine Officers and the RCPC.
- Review scheduling to ensure that adequate time is given to Canine Officers for veterinary and kenneling appointments.
- Ensure that Canine Officers are taking care of their detector dogs. Monitor and observe that Canine Officers: groom their dogs, routinely feed their dogs; maintain feeding schedules; follow health care schedules and maintain records; effectively handle injured or ill dogs; conduct daily health checks; routinely schedule dental care and health care visits; and monitor the service provided by veterinarians and their facilities.
- Provide support to RCPCs when conducting local detector dog teams validation testing

Maintaining Public Awareness and Communication

Local managers:

- Direct detector dog activities at their work locations and keep the RCPC informed.
- Work with Canine Officers before and during local presentations.

Handling Media Events

Local managers:

- Ensure that all media events are channelled through the RCPC.

Utilizing Detector Dogs

Local managers:

- Introduce Canine Officers to the work location (including kennels, port locations, etc.) and to PPQ.
- Facilitate understanding and support. Local managers may communicate to U.S. Customs, Immigration, Public Health, and port personnel about the operational procedures of detector dog teams.

Appendix B: Personnel
Local Managers (Supervisors, Port Directors, and SPHDs)

- Encourage and allow Canine Officers to work with RCPCs to develop different inspection techniques unique to local working environments. Detector dog teams' working areas are those for which they were trained by the NDDTC.

- Facilitate understanding and cooperation among the local work force by periodically assigning rotating PPQ officers or technicians to assist Canine Officers in inspecting flights during baggage inspection and allowing PPQ officers or technicians to assist in on-site training of detector dogs.

- Work with the RCPC to establish a work schedule for detector dog teams that will take advantage of international traffic (flights, mail, cargo) that best uses detector dogs, based on the results of feasibility studies, port records such as PPQ 212s, WADS, AQI monitoring data, pest risk, and country risk (high, medium, or low), and other port activities. With Canine Officers determine targeted flights within this time frame.

- Support training activities required of detector dog teams by allowing the time, the equipment, and the supplies necessary to maintain the proficiency of detector dogs.

- Provide vehicles to transport detector dogs from the kennel to the work site and other detector dog activities. In conjunction with RCPCs, ensure that Canine Officers are adhering to safety requirements of vehicles.

- Allow travel time to and from kenneling, training day, veterinary appointments, and periodic trips to purchase dog food, training supplies, etc.

Monitoring Reports and Documentation

Local managers:

- Ensure that Canine Officers properly document daily activities and prepare monthly reports.

- Ensure accuracy and timeliness (**by the 10th of the following month**) of monthly reports and monthly training records from Canine Officers.

- Send documentation to port directors and RCPCs.

Refer to **Appendix H**, **Reporting and Documentation**, for samples of the records maintained by Canine Officers.

Appendix B: Personnel
Regional Canine Program Coordinators (RCPCs)

Regional Canine Program Coordinators (RCPCs)

RCPCs:

- Manage detector dog activities for the region by establishing and implementing national goals and objectives for detector dog activities. Maintain systems to monitor regional activities, to identify problems, to provide solutions, and to report actions.

- Provide expert and technical advice about such topics as regional and national guidelines, limits of detector dog team utilization, and scheduling effectiveness.

- Ensure Canine Officers meet established national procedures in order to maintain a high level of proficiency in the detector dog teams of the region.

- Conduct site visits to gather information and provide support to Port Directors and Canine Officers in order to improve detector dog activities and procedures. Must visit work locations at least once a year; additional visits are scheduled as needed.

- Conduct new handler assessment. Upon completion, write assessment report. Send copies to the Port Director, the RPM, and NDDTC.

- Conduct replacement dog follow-up/assessment and upon completion, write assessment report.

- Conduct Canine Officer Supervisory Training (COST).

- Conduct feasibility studies within the region and recommend detector dog utilization to the appropriate manager.

- Provide technical advice about hiring Canine Officers, and serve as a member of selection panel, if requested, on a regional basis.

- Provide technical advice about selecting veterinarians for detector dog health care and choosing kennels.

- Collect, review, and evaluate monthly statistical and narrative reports from the detector dog teams within the region. Prepare annual reports for interested parties at the regional and national levels.

- Represent regional detector dog program issues at regional and national meetings, as requested.

- Recommend retirement of detector dogs in conjunction/ consultation with RPM and NDDPM.

- Participate in workforce planning and future trends and direction for detector dog activities by evaluating statistical analyses and visual observations; identifying and establishing trends, maintaining statistical records, and knowing the proficiency of detector dog teams.

Appendix B: Personnel
Regional Program Managers (RPMs)

- Assess new areas to work detector dog teams.
- Handle administrative concerns related to managing detector dog teams, such as purchasing field-supplied equipment, acquiring kenneling and veterinary services, arranging for logistics, and preparing justifications.
- Assist the NDDTC in procuring detector dogs.
- Identify and recommend training needs and curriculum changes for regional detector dog personnel and initiate training and/or secure and obtain approval for funds to initiate training.
- Ensure that the regional public awareness efforts effectively utilize detector dogs to convey the PPQ mission to the public. Duties related to public awareness are arranging for and conducting interviews and providing printed information and electronic media.
- Conduct validation tests on detector dog teams within the region.
- Work with local managers to schedule training for new detector dog teams.
- Serve as a member of a detector dog team when not performing regional duties to maintain proficiency. In this capacity, all roles and responsibilities of Canine Officers would apply to RCPCs.

Regional Program Managers (RPMs)

RPMs:

- Ensure and promote consistency in detector dog operations.
- Represent regional detector dog activities at regional and national meetings upon request.
- Supervise and direct RCPCs.
- Get input from RCPCs for national policy.
- Provide assistance to RCPCs in resolving detector dog issues (requests for regional meetings, requests for funding, budget,)
- Coordinate with RCPCs to identify training needs for regional detector dog personnel, secure funding, make decisions on what and how.
- Submit an annual report to the NDDPM.
- Consult/advise with the NDDPM regarding retirement of detector dogs.

National Detector Dog Instructors

The National Detector Dog Instructors:

- ◆ Procure detector dogs; perform a final temperament test and give final approval of selected dogs.

- ◆ Conduct training of detector dogs before Canine Officers arrive at NDDTC. This training includes protocol training on specific meat and fruit products (target odors) and nontarget odor training. Then conduct training of detector dog teams including: basic conditioning and commands, adaptation, and exposure to work location environments and detection work.

- ◆ Conduct Basic Canine Officer Training (BCOT).

- ◆ Support work locations that have detector dog teams by:
    - ❖ Providing constructive feedback to Canine Officers relating to training procedures and problems for the first 6 months after leaving NDDTC. (This timeframe may be extended by any party of interest.)
    - ❖ Providing technical assistance upon request.

- ◆ Maintain training supplies and equipment at NDDTC. This responsibility includes acquiring and maintaining:
    - ❖ Uncontaminated baggage (hard- and soft-sided), cardboard boxes, and paper bags
    - ❖ Target items (meat, fruits, and vegetables)
    - ❖ Nontarget items typical of flights to be encountered at the assigned work location of the detector dogs
    - ❖ Veterinary and kenneling services
    - ❖ Dog food, leashes and leads, collars, reward pouches, crates, etc.

- ◆ Develop and maintain training materials, such as instructor guides, handouts, exercises, tests, and evaluation instruments.

Animal Care Technicians at the NDDTC

Animal Care Technicians at the NDDTC:

- ◆ Care for, feed, and attend to the health and general well being of dogs.
- ◆ Keep track of detector dog health records, insuring vaccinations are kept up to date, and all medical issues are taken care of and documented.
- ◆ Maintain an adoption list and the adoption of dogs at NDDTC.
- ◆ Pick up and ship dogs to and from NDDTC.
- ◆ Take dogs to the veterinarian for routine visits as well as emergencies.
- ◆ Train new Canine Officers in kennel care and basic animal health care.
- ◆ Order all kennel supplies.
- ◆ Clean kennel facilities daily and maintain overall kennel safety.
- ◆ Bathe dogs.
- ◆ Assist in procuring dogs.
- ◆ Assist the National Detector Dog Instructors with placement of dogs with new Canine Officers.

Professional Development Center (PDC)

The Professional Development Center is responsible for providing leadership and administrative support to the National Detector Dog Training Center (NDDTC). The PDC supports the NDDTC in the following ways:

- ◆ Assists with the development of strategic training goals and objectives that are aligned with national detector dog program activities, policies, procedures, and operational guidelines
- ◆ Ensures that the NDDTC has the fiscal and human resources needed to accomplish its mission
- ◆ Provides administrative assistance with fiscal and human resource responsibilities such as status of funds, recruitment of personnel, procurement of outside services, etc.

National Detector Dog Training Center (NDDTC)

The National Detector Dog Training Center:

- ◆ Develops strategic training goals and objectives that are aligned with national detector dog program activities, policies, procedures, and operational guidelines
- ◆ Implements action plans for accomplishing strategic training goals and objectives
- ◆ Ensures that training provided is updated to reflect changes in national detector dog program policies, procedures, and operational guidelines
- ◆ Evaluates and assesses the effectiveness and efficiency of the training that is provided to national detector dog program personnel
- ◆ Provides technical assistance to Regional Canine Program Coordinators

National Detector Dog Program Manager (NDDPM)

In collaboration with the Regional Program Managers, the NDDPM:

- ◆ Ensures and promotes consistency in national detector dog program operations
- ◆ Provides assistance to the regions, upon request, in resolving detector dog program issues
- ◆ Represents detector dog program operations at national and international meetings and conference calls
- ◆ Formulates and writes national program policy for the detector dog program, with input from the RPMs and RCPCs
- ◆ Develops criteria/formulas/models/budget and resource justifications for planning and placement of detector dog teams and for presenting said proposals to PPQET
- ◆ Serves as the point of contact for requests for detector dog program assistance from other Federal agencies, national programs, and international sources and responds to their requests
- ◆ Works with the NDDTC to identify new training needs, supports present training initiatives, and assists in resource allocations
- ◆ Ensures that effective communication is a DDP norm

Appendix B: Personnel
National Detector Dog Program Manager (NDDPM)

Appendix C

History of Detector Dog Programs

Contents

Introduction **page-C-1-1**
History of Working Dogs **page-C-1-1**
History of Dogs as Scent Detectors **page-C-1-2**
History of Agricultural Detection by Dogs **page-C-1-3**
Scent Basics **page-C-1-5**
A Variety of Scent Duties **page-C-1-6**
Breed Selection Information **page-C-1-7**
   Beagles **page-C-1-7**
The National Detector Dog Training Center **page-C-1-8**

Introduction

When presenting media and public awareness events, detector dog handlers may find it useful to have interesting and accurate facts at their disposal about working dogs, the use of detector dogs, the breeds or types of dog they are handling, the scope of detector dog deployment, and some facts about scenting ability in dogs. The following sections provide plenty of information gleaned from reliable sources that can be used in a variety of public presentation formats.

History of Working Dogs

Dogs have a long history of working in partnership with human beings. Much of that history is bound up in warfare. The use of war dogs dates to centuries Before the Common Era (B.C.E.) as dogs had roles as warriors, guards and protectors in service to the Egyptians, Greeks, Assyrians, Persians and the Roman Empire. Roman legions deployed entire formations of armored attack dogs against enemy armies. Attila the Hun used mastiff-type dogs and Talbot hounds (ancestors of bloodhounds and beagles) as warriors in his campaigns and as sentries when his troops were encamped.

Dogs were used as sentries, guards, mascots, messengers, draft animals and scouts by armies worldwide. Our appreciation of their scenting ability was slow to develop. Perhaps the most famous early scenting dog was Barry, a Saint Bernard. From 1800 until 1812, Barry lived with the monks in a hospice in Saint Bernard Pass at an altitude of over 8000 feet. With his legendary little keg of brandy around his neck, Barry used his wonderful sense of smell to rescue over 40 persons during his career.

In the United States, during the French and Indian Wars in 1775, Benjamin Franklin recommended the use of dogs by the U.S. Army as a means of searching for marauders who were killing colonists and burning settlements near Boston, and in 1779, William McClay of Pennsylvania's Supreme Executive Council recommended using dogs to search for scalping parties. In 1835, the U.S. Army imported bloodhounds from Cuba with their handlers to use as man trackers in the swamps of western Florida and Louisana.

History of Dogs as Scent Detectors

As early as 1888, bloodhounds were employed by Scotland Yard for scent detection work in the "Jack the Ripper" case. Although the dogs did not figure prominently in the case, the British police and military continued to explore the use of dogs in scent work, and continued to train scent dogs, which they used in a limited capacity to detect land mines during World War I (WWI) and munitions caches in World War II (WWII). After WWII, the government of Finland started a land mine detection program based on the successful British program, and experimented with small breeds such as schnauzers and spaniels.

One of the first organized uses of scent detector dog units was by the Nazi army. They used tracker dogs to silently follow tracks of the British Special Air Services (SAS) officers who parachuted into Germany to collect intelligence prior to WWII. These dogs were trained to follow a given ground scent. The dogs sniffed a footprint, an article of clothing or a blood trail, and discriminated the specified trail among hundreds of other odors that had crossed it. German tracker dogs served dual duty; they located and attacked their quarry.

The British Army adopted the idea of using silent tracker dogs for location purposes only, and incorporated this training into their war dog program. In 1943, the British Army established "Recce Patrols," using human scouts and tracker dogs to locate the Japanese who were hiding on islands in the Pacific theatre.

Although the Metropolitan Police Department's Scotland Yard had trained dogs as substance scent detectors as early as the mid-1950's, the use of dogs for the detection of illegal substances, such as narcotics and explosives, began in earnest the 1960's. In 1968, the U.S. Department of Defense established a Military Working Dog Program at Lackland Air Force Base in San Antonio Texas and in 1971, began training detector dogs for drug interception duties on ships and aircraft returning from South Vietnam.

In late 1969, the U.S. Customs Service carried out a feasibility study on using dogs to detect narcotics and dangerous drugs. At Lackland Air Force Base, on April 1, 1970, U.S. Customs began an experimental narcotic detector dog training program, concentrating its efforts on training dogs to detect and respond to marijuana and hashish. Later that year, they expanded the targeted drugs to include cocaine and heroin. At the time, the success of training a drug detection dog on four odors was considered unlikely, and the Customs dogs gave the first example of the versatility of dogs in learning to discriminate several target odors.

At approximately the same time, the British Royal Army Veterinary Corps began training its own army dogs to detect drugs. They soon followed the drug detection program with explosives detection work to assist in quelling the strife in Northern Ireland. The U.S. followed suit and began training and deploying explosives detector dogs in 1973. By the mid-1970's government agencies throughout the world were using detector dogs for various specialized tasks.

In 1976, the U.S. Air Force started testing smaller breeds for detector tasks, including beagles and cocker spaniels. These small breeds had the advantage of easily searching close spaces that were inaccessible to the German shepherd dogs that had been used traditionally.

History of Agricultural Detection by Dogs

The Mexican government was the first to use dogs to detect agricultural quarantine items. In the late 1970's the United States Department of Agriculture (USDA) developed a similar program using dogs to search international mail and incoming passengers' baggage at international airports. The USDA began its training of agricultural detector dogs at Lackland Air Force Base. Until 1983, USDA used large breed dogs, such as Labrador retrievers and Australian heelers, and searches were conducted in non-public areas only.

The success of the large breeds created interest in a more visible canine detection program and USDA began to consider using small breeds to work in the presence of the public and in close proximity to international passengers. Beagles were selected by the USDA for the task as they are historically excellent scent hounds, and because the U.S. Navy had successfully deployed them as narcotic detector dogs. Also, due to their small, non-threatening size and appealing demeanor, they are extremely popular with the traveling public and the media. They are working advertisements for the mission of the USDA and the importance of agricultural quarantine work. Using beagles trained to indicate in a passive manner has had far-reaching effects on detection work world-wide.

Appendix C: History of Detector Dog Programs
History of Agricultural Detection by Dogs

In 1984, the USDA launched a novel pilot program with one beagle team at Los Angeles Airport. This was the beginning of the program we know as our "Beagle Brigade." At the time, the program was a major departure from any existing substance detection programs. The beagles were trained as they still are, to work close to the public at baggage carousels in international airports, and to respond passively, or to sit, when indicating the presence of agricultural quarantine material. Their reward is food for each correct response, which increases the intensity of their focus on task and the duration of their search time.

Beagles have long been the "workhorse" of the USDA program. Today the beagle has become known internationally as The Agricultural Detector Dog. Detector dog programs using beagles have been adopted in the U.S., Canada, Taiwan, Australia and New Zealand.

In 1997, the USDA responded to the threat of pests being introduced into the U.S. through land border crossings by deploying its first "Border Brigade" dogs. Beagles were used in a pilot project launched at Hidalgo, Texas. Beagle teams were placed at El Paso, Laredo and Hidalgo, Texas. Although the beagles were successful, by 2000, the agency returned to its use of large breeds due to the strenuous nature of performing vehicle searches. Currently there are "border dog" teams at both northern and southern borders with plans to expand the program.

In 2001, the USDA deployed its first cargo dogs in Texas, Washington State and California, again using larger breed dogs that are best suited to the stresses of searching containerized and palletized materials in warehouses and holding areas. These dogs are trained to use an active response to indicate the presence of specific agricultural materials in commercial shipments. We continue to evaluate the size and use of detector dogs in cargo situations.

In addition to the quarantine material detectors, the APHIS Wildlife Services (WS) has a detector dog program using terriers that are trained to sniff for brown tree snakes in aircraft, vehicles, household goods, and ships leaving Guam for snake-free areas like Hawaii, the Mariana Islands, and Saipan. Brown tree snakes have infested the island of Guam, causing the extinction of several native species of birds. To protect the rare flora and fauna in Hawaii, it is important to avoid accidentally introducing this destructive pest into the environment.

Appendix C: History of Detector Dog Programs
Scent Basics

Scent Basics

Dogs are wonderful odor detection devices because of their abilities to discriminate specific scents among complexes of many, overlapping scents. But there are other reasons why they make such good detectors of contraband materials. One way dogs are superior to gas-sensing machines as detection devices is that they can selectively locate odors. First, dogs sample air in an extremely efficient manner. Second, they are mobile, and can take their handlers directly to an odor source. As they move about, they can pick up the "thread" of an odor that interests them. By casting back and forth with their bodies and their heads, and by constantly taking tiny samples (sniffs), they compare odor concentrations and "calculate" the direction of increasing concentrations, following the molecular concentration gradient to its strongest point, or source. Fine-grained pinpointing of odor is achieved when the dog compares the strength of scent received in each of its mobile nostrils.

Dogs are extraordinary among mammals for their abilities in both reception and discrimination of odorants. Dogs can recognize the scent of table salt in 1:10,000,000 dilutions![1] Nearly 1/8th of the canine brain and more than half of the canine internal nose is committed to olfaction.[2] While human beings have a nasal epithelial area of 3-5 cm^2, dogs, in general, have an epithelial area ranging from 18-150 cm^2, and beagles, specifically, have an epithelium of about 75 cm^2. Beagles are just about average among dog breeds as far as being endowed with scenting equipment.[3] Another comparison, which unfortunately does not specifically list beagles, is that human beings have 5 million sensory cells; German shepherd dogs have 220 million sensory cells; dachshunds have 125 million; and fox terriers have 147 million. The number of sensory cells apparently correlates to some extent with the size of the dog. The larger area and proliferation of receptor cells most certainly enhances discriminatory ability.

---

1   Kaldenbach, Jan. 1998. K9 Scent detection. Detselig Enterprises, Alberta.
2   Syrotuck, W. G. 2000. Scent and the scenting dog. Barkleigh Productions, Inc., Mechanicsburg.
3   Albone, E. S. 1984. Mammalian semiochemistry investigation of chemical signals between mammals. Chichester, West Sussez: John Wiley and Sons Ltd.
    Dodd, G. H. and D. J. Squirrel. 1980. Structure and mechanism in the mammalian olfactory system. Symposia of the Zoological Society of London, 45:35-36.

The most remarkable aspect of the canine sense of smell is their ability to discriminate between complex mixtures of odors. Although dogs can detect odorants in quantities far lower than we can, it is the accuracy with which they can discriminate among odors that is the primary quality making them invaluable for odor detection. To illustrate this concept, also called "odor layering," consider this scenario: If you go into a kitchen where someone is cooking chili, you can smell chili. If a dog goes into the kitchen, it can smell the hamburger, the beans, the tomatoes, the garlic, each of the seasonings, and so on. The dog can separate each element of the chili into an individual layer or component of odorant.

Basically, if an odor is there, the dog can smell it. This is why odors cannot be effectively masked from detection by the canine nose simply by attempting to disguise them with stronger odors. Because dogs can detect minute quantities of odorants, and because they are considerably more capable of discriminating between the individual molecular combinations that identify odorants, attempts to "fool" detector dogs by packaging prohibited fruit or meat with other strong-smelling items are usually unsuccessful.

A Variety of Scent Duties

Today, detector dogs have many functions—they assist local, state and federal agencies in locating evidence, intercepting contraband and smuggled items, help police officers find criminals, lost children and the bodies of victims, are members of search-and-rescue teams, seek out land mines, search for live victims of earthquakes and other disasters, find evidence in arson investigations, detect explosive devices, poached abalone, and can even detect malignant skin growths. These jobs are mundane, though, when compared to some of the specialized uses of their scenting ability. The USDA has used dogs to detect screwworm infestations in cattle and the presence of brown tree snakes in aircraft. Australian shepherds have been used to indicate when cows are fertile so that farmers can breed them at the right time during their short estrus. Beagles are increasingly used to precisely locate termites in buildings to avoid the necessity of treating the entire structure. U.S. Customs has further expanded its war on drugs using dogs to indicate substantial amounts of currency that may be associated with a drug transaction. Dogs have been used to perform ecological studies of wildlife by indicating on the scats of specific animals, demonstrating that certain species are present in an area under study. Even with work on an "artificial nose" for detection of substances continuing, dogs are being used with increasing frequency to detect a variety of substances.

Breed Selection Information

All dogs bring superior scenting abilities to the table. Limitations on the endurance required to perform scent work, however, eliminates the extremely short-nosed breeds (e.g., pugs, Boston terriers, Pekingese, etc.) as working scent detection dogs. The German shepherd dog was traditionally the breed of choice due to its long history of development for working ability. However, it, and other excellent large working breeds, such as Rottweilers, Belgian malinois, and Bouvier des Flandres have also been bred for protective devotion to their handlers which can create focus difficulties for a dog team deployed in highly public areas. Because the USDA is a regulatory agency that stresses voluntary compliance, breeds that are less likely to be associated with protective, military and law enforcement functions enhance the image of the agency.

Beagles

As members of the hound group, beagles have been bred as pack hunters of small game for centuries. They were first recognized by the American Kennel Club (AKC) in 1885. The pack orientation of beagles makes them very sociable. They are extremely curious, gentle, and are among the least aggressive of all dog breeds. They have excellent memories and are very active. Their focus on interesting scent and love of food is legendary, all of which qualities make them ideal as highly visible USDA detector dogs in close contact with the public.

The National Detector Dog Training Center

After the early training at Lackland Air Force Base, from 1986 to 1987, APHIS opened 3 regional training centers, in New York, Miami and San Francisco, and in 1988, began training its own teams. The major contributors to the fledgling program were Douglas R. Ladner, PPQ Senior Staff Officer, and PPQ Officers Mike Simon, Mel Robles, Cal Brannaka, and Hal Fingerman. In October, 1997, the National Detector Dog Training Center (NDDTC) was opened in Orlando, FL, unifying the 3 regions. The NDDTC is located on almost 2 acres of land, the buildings occupying 7,800 square feet. The facility includes 30 kennels, 5 quarantine runs, mail and passenger training areas, and classrooms. An alternate site was opened in 2000 for use by trainers for preparing the dogs for classes. All PPQ Canine Officers and detector dogs complete basic training at the NDDTC. Additionally, more advanced training and training for other countries is performed at the training center. NDDTC trainers have provided expertise and training to agriculture officials in other countries starting up their own detector dog programs (e.g., Taiwan, Canada, Australia, New Zealand, Guatemala, Mexico, and South Korea). NDDTC trainers contribute a tremendous diversity of experience and training techniques to the success of the program.

The mission of USDA's National Detector Dog Training Center is to "provide a positive learning environment to train detector dogs and officers as teams to safeguard American agriculture" and to "develop innovative ways for these teams to prevent pests and agricultural diseases from entering the United States through airports, international borders, mail facilities and cargo areas."

Appendix D

Equipment

Contents

Introduction **page-D-1-1**
Collars **page-D-1-1**
   Elizabethan Collar **page-D-1-2**
   Martingale Collar **page-D-1-3**
   Regular Nylon Collar **page-D-1-4**
   Slip Collar **page-D-1-4**
     How to Place a Slip Collar on a Dog **page-D-1-5**
   Leather Collar **page-D-1-6**
   Harnesses **page-D-1-6**
   Crates/Portable Kennels **page-D-1-7**
Crates/Portable Kennels **page-D-1-7**
   Crate Pads **page-D-1-8**
First Aid Kit **page-D-1-8**
Trauma kit **page-D-1-10**
Grooming Kit **page-D-1-12**
Jackets **page-D-1-13**
Leashes (Regular and Retractable) **page-D-1-13**
Refrigerators **page-D-1-14**
Reward Pouch **page-D-1-14**
Suitcases, Boxes, and Contents **page-D-1-15**
Vehicles **page-D-1-16**

Introduction

In this appendix are descriptions and explanations of the equipment and supplies listed under the **Equipment** section and those mentioned in other sections of the manual. The items are listed alphabetically. **Purchase this equipment as soon as possible.**

Collars

Collars for detector dogs should be nylon. They must be black or forest green. See **Figure D-1-1**. Collars that buckle or snap adjust to a fixed size. They are designed not to tighten around a dog's neck if it is on a loose lead, but will tighten if the dog pulls. The primary consideration when selecting a collar for a detector dog is the dog's safety. The collar must fit securely to prevent the dog from escaping from the collar or slipping out of it. It should not fit so tightly as to interfere with the dog's comfort while working. Generally, only one or two fingers should fit in the space between the dog's neck and the collar when the collar is in its normal position.

Appendix D: Equipment
Collars

All slip collars must be removed for the detector dog when it is kenneled, and special care must be taken to remove slip collars whenever the dog is not being directly observed or supervised.

See additional information that follows about martingale, slip, leather, and nylon collars, as well as directions on how to make an emergency Elizabethan collar.

FIGURE D-1-1: Collars Worn by Detector Dogs

Elizabethan Collar

An Elizabethan collar is used to prevent a dog from licking or chewing itself, or pawing at and rubbing its face after an injury before you can get it to the veterinarian. See *Figure D-1-2*. You can make an emergency Elizabethan collar by cutting the bottom out of a plastic planting pot. It should be large enough to just slip over the dog's head. Cut four holes in the pot and slip a strip of gauze through each hole. Secure the emergency Elizabethan collar to the dog's collar, using the gauze strips and tape. If the dog is not wearing a collar, create a makeshift collar with a piece of gauze tied loosely around the neck.

Appendix D: Equipment
Collars

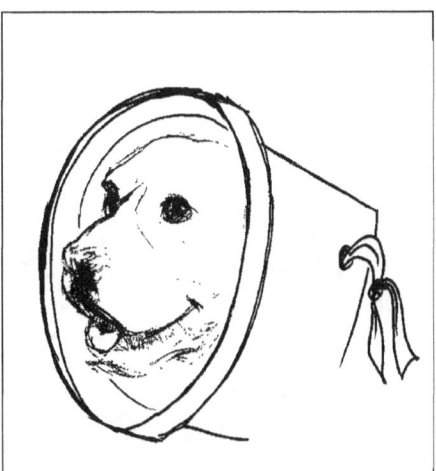

FIGURE D-1-2: Emergency Elizabethan Collar Made from a Plastic Planting Pot

Martingale Collar

A Martingale collar is a fixed or modified slip collar. It will stop constricting at a point that you adjust. The tightest point should be somewhat smaller than the broadest part of the dog's head to prevent escape. The collar fits loosely. It may fit more than two fingers between the neck and collar, but not loosely enough to hang slackly over the dog's chest or allow the head to be pulled out of it (see *Figure D-1-3*). Martingales may be used during passenger clearance.

FIGURE D-1-3: Martingale Collar

Regular Nylon Collar

Nylon collars must be black, forest green, or brown. Inspect nylon collars for excessive fraying. Check the plastic snaps for splits, cracks, or breaks. To clean nylon collars, rinse them in soapy water or wash them in a washing machine (see *Figure D-1-4*).

FIGURE D-1-4: Regular Nylon Collar

Slip Collar

A slip collar is designed to tighten around the neck of the dog when tension is applied on the leash. Slip collars can be metal chain, leather, or nylon (see *Figure D-1-5*). Never use slip collars during passenger clearance.

 Only use these collars during training exercises. **Never use a slip collar during passenger clearance.**

Appendix D: Equipment
Collars

FIGURE D-1-5: Slip Collar

How to Place a Slip Collar on a Dog

1. Put your leash clip on the top ring. Hold the slip collar in a vertical position, and drop the chain through the bottom ring.

2. With your dog in front of you, slip the collar over the dog's head. The ring with the leash clip should be the ring that moves. To check it, gently pull the lead and release. The collar should tighten and immediately loosen (see *Figure D-1-6*).

FIGURE D-1-6: Correct Way to Place a Slip Collar

3. If the lead does not loosen, it is on upside down. Take the collar off and turn it over (see **Figure D-1-7**).

FIGURE D-1-7: Incorrect Way to Place a Slip Collar

Leather Collar

Give regular care to the metal buckle and "D" ring on the collar to avoid rust along the edges and where metal joins the leather. Also, apply a leather conditioner or saddle soap to the outside of the collar.

Harnesses

Harnesses are another approved form of restraint for special purposes. You **must get approval** from your RCPC before purchasing a harness.

For an example of a harness, see **Figure D-1-8** and **Figure D-1-9**.

FIGURE D-1-8: Harness **FIGURE D-1-9: Harness Placement**

 Get approval from your RCPC before purchasing a leather collar or harness.

Crates/Portable Kennels

There are two types of enclosures used to house a detector dog while working, and they are considered a detector dog's secondary residence. One is a wire crate, and the other is a portable kennel. See *Figure D-1-10* and *Figure D-1-11*. Wire crates are recommended for transporting detector dogs in vehicles, allowing for greater air circulation. Portable kennels are used to ship detector dogs. Wire crates or portable kennels can be used to house detector dogs between flights.

The approximate size of a crate or portable kennel must be large enough for the dog to comfortably stand up, lie down, and turn around. Crates and portable kennels must have a solid floor. See the definition of secondary residence in the *Glossary*.

 Never use a portable kennel if the structural integrity of the kennel has been compromised in any way.

FIGURE D-1-10: Wire Crates Used for Detector Dogs

Appendix D: Equipment
First Aid Kit

FIGURE D-1-11: Portable Kennels Used for Detector Dogs

Crate Pads

Detector dogs need crate pads to reduce injuries. Padding in the crate keeps the dog comfortable, prevents elbow calluses, and helps the dog regulate its temperature. Crate pads may be made of "sheepskin," plush, or smooth fabric, may be covered foam mattresses, or constructed of any other sanitary and washable material. Always inspect crate pads for signs of chewing, as ingestion of padding material can be unhealthy for the dog (i.e., in some cases, can create intestinal obstruction).

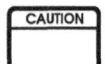 Some dogs chew their crate pads. Be observant and if your dog chews its pad, give the dog chew toys. If that does not work, remove the crate pad.

First Aid Kit

Canine officers are responsible for maintaining a fully-stocked first aid kit and ensuring it is accessible for first aid and emergency care of their detector dogs. Contact your local approved veterinarian to replace used or expired items. New Canine Officers receive a first aid kit during BCOT at NDDTC. Following is a list of items in the first aid kit (see *Figure D-1-12*).

- Adhesive tape (coflex)
- Alcohol prep pads
- Antibiotic ointment
- Cold pack

- Cotton balls, rolls, swabs
- Eye wash
- Gauze bandage rolls
- Gauze pads, sterile non-stick
- Gloves
- Iodine prep solution
- Magnifying glass
- Providone—Iodine ointment
- Scissors, blunt-tipped
- Thermometer
- Tweezers
- Blanket (not in the kit)
- Muzzle (not in the kit)
- Soft, nylon rope for emergency leash (not in the kit)

Items that are not in the kit but are useful to stock:

- Blanket
- Muzzle
- Soft nylon rope for emergency leash

FIGURE D-1-12: First Aid Kit Supplied by NDDTC

Appendix D: Equipment
Trauma kit

Trauma kit

Each port location should have access to a trauma kit. Items include the following (see **Figure D-1-13**):

- Adhesive tape 1"x2.5 Yds.
- Adhesive tape 1"x5 Yds.
- Alcohol wipes
- Ammonia inhalant
- Aspirin, buffered 2 Pks.
- Benadryl 25 mg Tabs 2 Pks.
- Benzalkonium chloride wipes
- Case, nylon
- Cast padding 3"x 4Yds.
- Co-Flex 2"
- Co-Flex 4"
- Cold pack 5x6
- Combine dressing 8x 7.5 Yds.
- Combine pad 5 x 9 Yds.
- Cotton applicators, non-sterile 3"
- Cotton applicators, sterile 6" 2 Pks.
- Elastic bandage 3" No Latex
- Emergency care and CPR card
- EMT gel
- Eye and skin wash 4 oz.
- Forceps
- Gauze pad 2" x 2"
- Gauze pad 3" x 3"
- Gauze pad 4" x 4"
- Gauze roll 2" x 4Yds.
- Gauze roll 4" x 4 Yds.
- Gauze trauma pad 5" x 9"
- Gloves, vinyl
- Green soap towlettes
- Handwipe

Appendix D: Equipment
Trauma kit

- Hydrocortisone cream 1gm
- Insect sting swabs
- Loperamide caplettes
- Non-adhering pad 2"x 3"
- Non-adhering pad 3" x 4"
- P.A.W.S Wipe and Dry
- Peroxide
- Petrolatum white jelly
- Petscope
- Pill gun
- Plastic tweezers 4"
- PVP iodine 1 fl oz.
- PVP iodine swabs
- Razor
- Scalpel
- Scissors, paramedic
- Scissors, small
- Skin stapler
- Space blanket
- Stethoscope
- Stockinette 3' wide
- Syptic powder
- Syringe, 60cc irrigation
- Syringe bulb
- Thermometer
- Tick remover
- Tongue depressors
- Tourniquet, powder free, latex free
- Triple antibiotic, 1 gm

Appendix D: Equipment
Grooming Kit

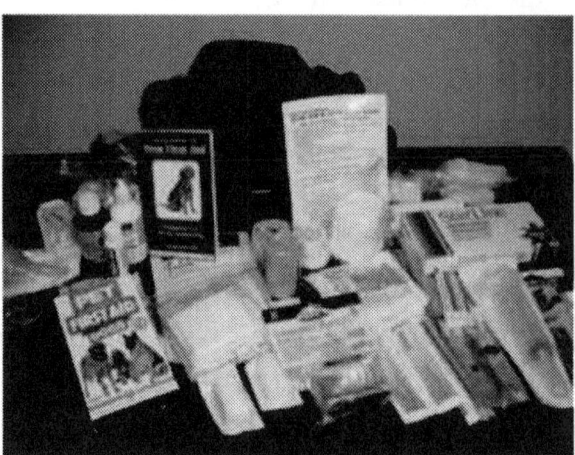

FIGURE D-1-13: Trauma Kit

Grooming Kit

Each grooming kit contains a brush, comb, nail clippers, and other similar items used to remove shedding hair and to maintain nails. Which tools to use will depend on each dog's type of hair and skin (see *Figure D-1-14*). Also needed are an ear cleaner, shampoo, and styptic powder.

Some dogs will have specific needs that may require different grooming supplies. Consult with your RCPC for more information.

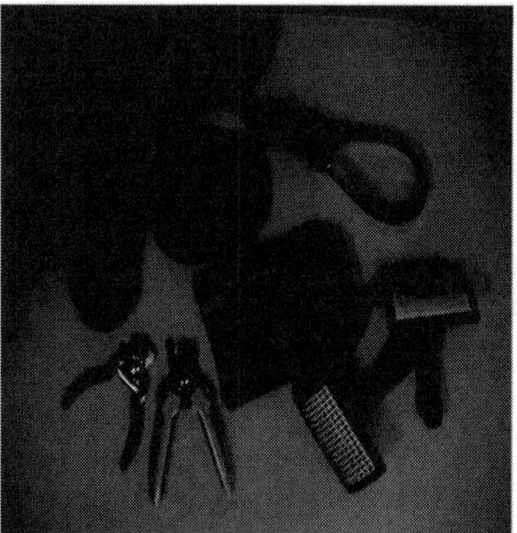

FIGURE D-1-14: A Variety of Grooming Tools

Appendix D: Equipment
Weight Scale

Weight Scale

Each work unit should have access to a scale (either at the kennel or work location) to help maintain the detector dog's optimum weight.

Jackets

The detector dog's uniform is an official, green jacket that must be worn when working passenger clearance and when present at an official event. Jackets are issued at NDDTC during BCOT. If additional jackets are needed, contact your RCPC.

Wash and maintain the jacket, collar, and bedding at least once a month.

Important

Each detector dog should have a jacket to be used only for media presentations. The jacket should **not** have the reflective strip on it.

Leashes (Regular and Retractable)

A regular leash is used while working a detector dog, and a retractable leash may be used to exercise a detector dog. The detector dog should be on a leash when it is not kenneled or in a crate (see **Figure D-1-15**).

A regular leash may be nylon, cotton, or leather. Leashes must be black, forest green, or brown. Their length should not exceed 6'.

Never use a retractable lead while working the dog.

Important

Inspect the metal snap on a leash daily, because the spring in the snap can become faulty. Strain will cause the curved portion of the snap to pull out of shape. Keep the snap free of rust. Inspect the leash for cuts, wear, and tear daily, and oil it frequently.

To condition leather, use saddle soap or a leather conditioner. Take care when conditioning leather. Excessive leather conditioner or saddle soap can cause the leash to stretch until it becomes unsafe. To clean nylon or cotton leashes, rinse them in soapy water or wash them in a washing machine.

A retractable leash may be used to exercise a detector dog in uncrowded areas. This allows the dog to run and exercise while secured by a leash.

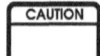 Hold on to the leash at all times. When exercising the dog, never drop the retractable lead because it will fly back and hit the dog.

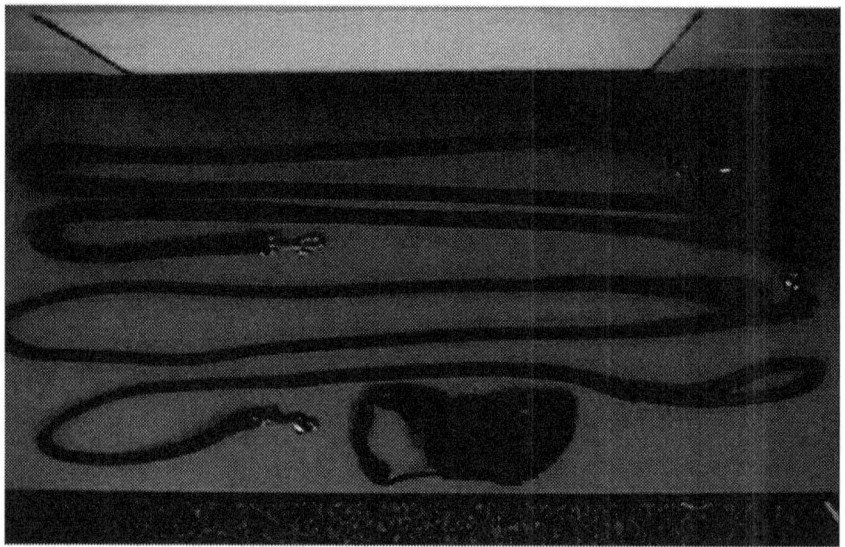

FIGURE D-1-15: Leashes (Regular and Retractable)

Refrigerators

A refrigerator is required to store target and nontarget items used for training. It is recommended that there be 2 refrigerators—1 to store target items, and the other to store nontarget items.

Additionally, air-tight containers are required to store items in the refrigerators. Mark the containers, and only store items in each container as marked. For example, store citrus in one container, apples in another, cheeses in another, breads in another, etc.

Reward Pouch

A reward pouch is required to store treats or food rewards for a detector dog. Reward pouches are supplied by NDDTC during BCOT. Canine Officers wear the reward pouches on their belts.

Appendix D: Equipment
Suitcases, Boxes, and Contents

If a pouch needs to be replaced, contact your RCPC. Note that fanny packs work well. They must be black and plain with no fancy stitching or color, and can be leather, nylon, or vinyl (see *Figure D-1-16*).

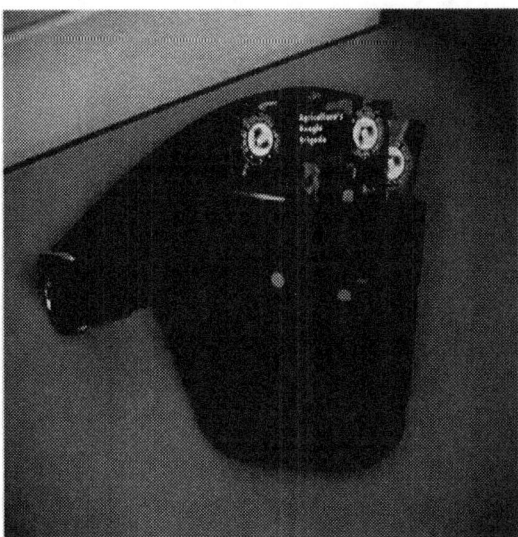

FIGURE D-1-16: Reward Pouch

Suitcases, Boxes, and Contents

A supply of hard suitcases, soft suitcases, handbags, and boxes, along with clothes, shoes, and target and nontarget items to fill them are required for proficiency training. Suitcases, handbags, and boxes used to hold target items and those used to hold nontarget items need to be stored separately. Clothes and other items are needed to fill the suitcases.

Suitcases and handbags—minimum of 50 hard and soft cases in a variety of sizes; including 50 percent handbags, backpacks, purses; excluding target cases (see *Appendix D* under **Suitcases, Boxes, and Contents** for more information). Additionally, maintain 25 boxes.

Important
Be aware of any second-hand suitcases, bags, and boxes you may procure from thrift shops, etc. These secondhand items may have a residual odor because of what they were used for in the past, such as holding food. If you do not know the origin of a suitcase, bag, or box and the dog continues to respond to it, then throw it away.

Important
Canine officers are recreating passenger baggage for training scenarios. Therefore, the suitcases and handbags should be filled with anything encountered on the passenger clearance floor.

Appendix D: Equipment
Vehicles

Maintaining this supply ensures the highest level of stimulation when training detector dogs, which in turn helps to maintain a high level of proficiency.

Vehicles

Safe vehicles (passenger vans are ideal) and equipment are required to transport detector dogs. Local management is responsible for ensuring that a Canine Officer adheres to the safety requirements of a vehicle. Vehicle maintenance must be performed according to GSA and APHIS standards. Following are the safety requirements of vehicles and equipment used to transport detector dogs. The list is periodically reviewed and approved by the APHIS National Safety and Health Committee.

- Properly working air conditioner and heater, as dictated by climate. Air conditioners and heaters are mandatory. Rear air conditioning is highly recommended, and may be mandatory in some climates to maintain the appropriate temperature for transporting the dog. **Do not use a vehicle without air conditioning to transport a dog if the temperature is above 85°F.** Before placing a detector dog in a vehicle, allow the vehicle to cool down (or warm up).
- Tinted windows that are fully functional. Tinted windows reduce solar convection.
- Alternative power source such as a battery to operate equipment, such as phones or fans.
- Wire crate to house the dog in a vehicle.
- First aid kit for dogs and humans. Kits are supplied by NDDTC to new canine officers. The RCPCs maintain the kits by routinely monitoring supplies and replacing expired and used items. See *First Aid Kit* for a list of included items.
- Fire extinguisher
- Communication device for emergency use, as dictated by the region or work location depending on the local circumstances and distances between the kennel and work sites (cellular phone, etc.).
- Emergency kit, including reflective triangles

Appendix E

Shipping and Daily Transporting Detector Dogs

Contents

Daily Transporting Detector Dogs **page-E-1-1**
Shipping Detector Dogs **page-E-1-2**
  Before Shipping **page-E-1-2**
  During Shipping **page-E-1-3**
  After Shipping **page-E-1-4**

Daily Transporting Detector Dogs

PPQ is responsible for providing vehicles to transport detector dogs from the kennels to work sites, to other canine activities, and to return to the kennels. It is the responsibility of canine officers to maintain the vehicles.

Vehicles should be large enough to transport detector dogs in traveling crates along with the support equipment and sufficient space for air circulation. The best vehicles for transportation are passenger vans or utility vehicles.

Vehicles should be assigned permanently to detector dog activities, as long as these activities are being performed. At work locations where there are more than one detector dog team (a canine officer and a detector dog), a vehicle should be assigned to each team.

Parking should be designated in official areas as close as possible to the inspection area to facilitate unloading and loading of the detector dogs and support equipment. Vehicles may be parked at kennels if there is a time savings and the security of the vehicles justifies the action. Port directors should make the final decision.

Refer to **Appendix D**, **Equipment**, under **Vehicles** for a list of safety requirements for vehicles.

Appendix E: Shipping and Daily Transporting Detector Dogs
Shipping Detector Dogs

Shipping Detector Dogs

Detector dogs can be shipped to their assigned work locations or to temporary locations to participate in temporary duty (TDY) assignments or advanced training. While being transported, detector dogs are protected by the AWA. Therefore, it is critical when shipping detector dogs, that they are transported in full compliance with the AWA.

It is the responsibility of canine officers to ensure that all safety conditions are met for shipping detector dogs. Refer to **Appendix D, Equipment**, for safety requirements for crates and portable kennels.

Not all airlines transport dogs.

During certain times of the year, airlines impose shipping restrictions due to weather conditions.

Before Shipping

1. Within 10 days of travel, get a health certificate and letter of acclimation from the dog's veterinarian. Ask the veterinarian to provide any required vaccination or treatment. The letter of acclimation is a statement on the veterinarian's letterhead attesting that the dog is healthy and capable of becoming accustomed to a new environment or situation, such as air travel.

   For trips to Hawaii, U.S. Territories, and certain foreign governments, you must check on any quarantine or health requirement needed for arriving animals at least four weeks in advance. Use **Table E-1-1** to find out where to check.

TABLE E-1-1: Where to Check About Needed Quarantine or Health Requirements for Arriving Animals

For information on requirements for:	Contact:
Hawaii[1]	Your State veterinarian's office or local Veterinary Services
U.S. territories: Puerto Rico, U.S. Virgin Islands, and Guam	
Foreign countries	The appropriate embassy, governmental agency, consulate, or NDDTC

1 Consult Hawaii's Department of Agriculture at 808-483-7151 for information about pre- and post-arrival requirements, quarantine station procedures, policies, rules, operations, and fees.

2. When making airline reservations, do the following:

   A. Use direct flights, whenever possible, to avoid accidental transfers or delays.

B. Travel on the same flight as the detector dog, whenever possible.

C. A detector dog traveling by air without an accompanying handler will do so by using a Government bill of lading or a Government issued credit card.

D. In the summer, choose early morning or late evening flights to avoid temperature extremes that may affect the dog.

3. Contact the airline to determine if additional requirements are to be met. Arrange for the detector dog to travel in the cargo compartment of the plane. Reconfirm with the airline 24–48 hours before departure that you are transporting a detector dog.

Airlines reserve the right to refuse to carry an animal for any reason.

4. The portable kennel must meet airline standards and requirements. Call the airlines or consult the AWA for the minimum standards for identification, sanitation, size, strength, and ventilation. The size of the portable kennel must be big enough for the dog to comfortably stand up, lie down, and turn around.

Never use a portable kennel to ship a detector dog if the structural integrity of the kennel has been compromised in any way.

During Shipping

1. Present the detector dog for transport at the airline cargo terminal no more than 2 hours before flight time.

A. If you are traveling with the dog, verify with the flight crew that an animal is on board the aircraft prior to departure.

B. If you are not traveling with the dog, call the party receiving the dog and let them know the dog was sent.

2. Carry a leash with you so that you may walk the dog before check-in and after arrival. Do not place the leash inside the kennel, and do not attach it to the outside of the kennel (the dog may chew it).

3. Carry a current photograph of the detector dog. If the dog is accidentally lost, having a current photograph will make the search easier.

Appendix E: Shipping and Daily Transporting Detector Dogs
Shipping Detector Dogs

After Shipping

1. If you are not traveling with the dog, ask the receiving party to call when the dog arrives.
2. If the dog should turn up missing during transport, immediately speak to airline personnel.
3. If the dog is not found, proceed with the following steps:
   A. Contact animal control agencies and humane societies in the local and surrounding areas. Check with them daily.
   B. Contact APHIS, Animal Care regional office closest to where the detector dog was lost. (See **Appendix A**, **APHIS Contacts**, for a list of regional offices in Animal Care.)
   C. Contact your RCPC and port director.
   D. Provide descriptions and photographs to the airline, local animal control agencies, and humane societies. Help can also be sought from radio stations. Leave your telephone numbers and addresses with all these people or businesses should you have to return to your work location.

Appendix F

Weight Rating

The weight of an animal depends on several factors. Many veterinary nutritionists have illustrated weight ranges for dogs from very thin to obese. Keep in mind that the age of a dog will factor in to the ideal weight; as well as the body changes that occur when aging from adolescence to maturity. Like people, as a dog ages and matures, its body shape changes. For example, at about 2–3 years old a dog's rib cage springs out and its chest drops. Some dog's appearance changes considerably when this happens. Also, weight gain may accompany this change.

Use **Table F-1-1** on the following page to match the physical characteristics with your detector dog to determine if it is very thin, ideal, or obese. If the weight of your detector dog is other than ideal, consult with a veterinarian about a prescribed diet.

Appendix F: Weight Rating

TABLE F-1-1: Determine the Weight Range of Your Dog

If the physical characteristics of your detector dog are:	Then consider your dog at a weight range of:
◆ Ribs are easily felt with no fat cover ◆ At the base of the tail, the bones are raised with no fat between the skin and bone ◆ From a side view, there is a severe abdominal tuck ◆ From an overhead view, there is an hour-glass shape	Very Thin. Consult with a veterinarian about a prescribed diet.
◆ Ribs are easily felt with slight fat cover ◆ At the base of the tail, there is a smooth shape where the bones can be felt under a layer of fat ◆ From a side view, there is an abdominal tuck ◆ From an overhead view, there is a well-proportioned lumbar waist	Ideal
◆ Ribs are difficult to feel and are under a thick layer of fat ◆ At the base of the tail, it is thick and difficult to feel the bones under a thick layer of fat ◆ From a side view, fat hangs from the abdomen and there is no waist ◆ From an overhead view, it is broad with no shapeliness	Obese. Consult with a veterinarian about a prescribed diet.

Appendix G

Expressing Anal Glands

Expressing Anal Glands

All dogs have a pair of sac-like anal glands located under the tail, just inside each side of the anus (*Figure G-1-1*). The sacs hold a thick, foul smelling, brownish glandular fluid. The fluid is naturally secreted in minute quantities as the dog defecates. Although domestic dogs rarely release these glandular secretions for scent marking, they can, in some cases, be voluntarily emptied when the dog is extremely alarmed, struggling, or as a self defense mechanism. If the dog produces large quantities of the secretion, or if the ducts leading to the anus become blocked, the glands can become impacted which can lead to the formation of an abscess. An abscess, if untreated, may rupture through the skin and will require veterinary treatment. Dogs with impacted anal glands may "scoot" their rear ends on the ground to relieve the pressure in the glands, or they may lick the anal area excessively or "flag" the tail to one side.

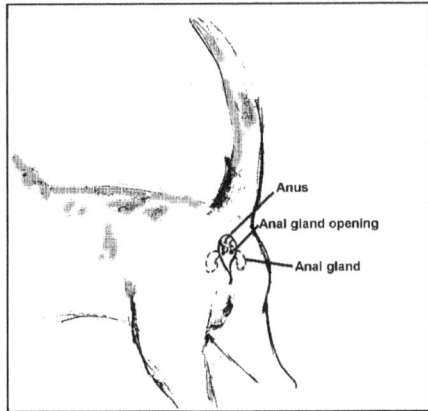

FIGURE G-1-1: Location of Anal Glands

The care required of the anal glands is highly individualized. If you see any of the above signs, take the dog to a veterinarian or express the anal glands using the following steps:

Step 1:
Secure the dog, as some dogs resent having their anal glands expressed, and most try to investigate the process once it has begun.

Step 2:
Grasp the base of the dog's tail with one hand, gently raise the tail to expose and slightly tighten the skin over the anus.

Appendix G: Expressing Anal Glands
Expressing Anal Glands

Step 3:
Place the thumb and forefinger of either hand on each side of the anus. You may feel two "pea-like" lumps, which are the glands.

Step 4:
Press the thumb and forefinger in and together, then squeeze, sliding fingers out toward the anal opening. Foul, fishy-smelling fluid may drip or squirt from the anus. If the secretion is white, bloody, or contains pus, you should contact your veterinarian.

Step 5:
Wipe the anal area to remove the fluid. To ensure that you have removed all traces of the fluid, use a damp, soapy cloth and a mild antibacterial soap.

Appendix H

Reporting and Documentation

Contents

Introduction **page-H-1-1**
Statistical Summary **page-H-1-2**
  Purpose **page-H-1-2**
  Standard Statistics Recorded Electronically **page-H-1-2**
Narrative Report **page-H-1-3**
  Purpose **page-H-1-3**
Canine Pest Identification Log **page-H-1-3**
  Purpose **page-H-1-3**
Distribution of Operational Reports **page-H-1-5**
Baggage Information Data (PPQ Form 277) **page-H-1-5**
  Purpose **page-H-1-5**
Training Records **page-H-1-6**
  Purpose **page-H-1-6**
  Completing the Training Record Form **page-H-1-6**
  Example of a Completed Training Record **page-H-1-8**
Health Care Records **page-H-1-10**
  Purpose **page-H-1-10**
  When and How to Maintain **page-H-1-10**
  Distribution **page-H-1-10**
CBP Directive No. 3340-025B: Commissioner's Situation Room Reporting **page-H-1-10**
  1: Purpose **page-H-1-10**
  2: Authority/Background **page-H-1-11**
  3: Scope/Definition **page-H-1-11**
  4: Reporting Procedure **page-H-1-11**
  5: Written Response **page-H-1-11**
  6: Responsibilities **page-H-1-11**
  7: Reportable Items **page-H-1-12**
  8: Controlled Deliveries **page-H-1-15**
  9: Requests for Information **page-H-1-15**
  Instructions for Completing Significant Incident Report **page-H-1-16**

Introduction

The worksheets, forms, and documents mentioned in this manual provide the required statistical information to monitor accomplishments of detector dog activities on a national level.

Regional Directors, RCPCs, or other local managers may desire additional information to be collected, recorded, and submitted by Canine Officers. The worksheets and directions for local or regional information will be provided by RCPCs or other local managers.

Appendix H: Reporting and Documentation
Statistical Summary

Statistical Summary

Purpose

Provide statistical information to monitor accomplishments of detector dog activities at all management levels.

All Canine Officers are responsible for recording detector dog activities, using nationally approved worksheets. Other worksheets may be used to collect port-specific information when required by local management.

Important

Canine officers who work clearing mail material or border cargo use different worksheets to record their activities, because they need to record different statistics than what is needed to analyze proficiency in clearing passengers, pit baggage, and crew of international aircraft.

Standard Statistics Recorded Electronically

E-mail the following standard statistics as follows to the Port Director and RCPC:

1. Number of days the detector dog team worked.
2. Number of flights screened.
3. Total passenger count from flight declarations.
4. Number of passengers screened.
5. Total number of responses—includes false and positive responses.
6. Number of positive responses—those that result in a seizure of prohibited agricultural items, and of similar but enterable agricultural items that are not nontarget material. Positive responses also include residual odors—when a target item was present in the container but is no longer there within a reasonable period.
7. Number of responses with seizures—those that result in a seizure of prohibited agricultural items. The number of responses with seizure column equals the sum of the total of handbags and pit bags seizures.
8. Number of quarantine material interceptions (QMIs)—plant (vegetative materials including fruits, vegetables, and plants), animal (products of animal origin including pork, beef, poultry, milk products, skins, feathers, manure, etc.), and weight of the animal QMIs (in kilograms).

9. Number of seizures—responses with seizures of prohibited agricultural items in handbags (baggage handcarried on the aircraft by the passengers), pitbags (checked-in luggage, suitcases, duffel bags, boxes—wooden or cardboard, or in any other large receptacle), and crew. Categorize crew seizures as either hand bag or pit bag.
10. Number of penalties, and total of fines (optional).
11. Percentage of accurate responses (calculation).

Important

Monthly reports are due by the 10th of the following month to the Port Director and the RCPC.

For the distribution of the statistical summary, see *"Distribution of Operational Reports"* on **page-H-1-5**.

Narrative Report

Purpose

The report provides information that cannot be recorded and conveyed in statistical summaries. Canine officers are responsible for writing monthly and narrative reports to accompany the statistical summaries that go to their supervisors and RCPC.

A narrative report may include the following topics. This list is not all inclusive.

- Demonstrations and other public awareness activities
- Reportable pest interceptions on the Canine Pest Identification Log for significant pests (see *Figure H-1-1*)
- Update on maintenance of medical records
- Working highlights
- Completed training

For the distribution of the narrative report, see *"Distribution of Operational Reports"* on **page-H-1-5**.

Canine Pest Identification Log

Purpose

The log provides information about significant or actionable pests intercepted by detector dog teams. This information accompanies the statistical summaries and the narrative reports submitted by Canine Officers.

Appendix H: Reporting and Documentation
Canine Pest Identification Log

Refer to **Figure H-1-1** for a sample of a Canine Pest Identification Log that Canine Officers can use to record the necessary information about significant or actionable pests intercepted by their detector dog. The information includes the following:

1. Country of origin—Record the origin of the host material, which may or may not be the origin of the flight.
2. Host—Record the scientific name of the host material.
3. Pest identification—Order, family, genus, species. Record the identification down to the most specific name.
4. PPQ 309 number—Number of the PPQ Form 309 submitted with the interception.

For the distribution of the canine interception log, see *"Distribution of Operational Reports"* on **page-H-1-5**.

CANINE PEST IDENTIFICATION LOG

Work Location_____   Month/Year_____

COUNTRY OF ORIGIN	HOST	PEST IDENTIFICATION	PPQ 309 NUMBER

FIGURE H-1-1: Sample of a Canine Pest Identification Log

Appendix H: Reporting and Documentation
Distribution of Operational Reports

Distribution of Operational Reports

Canine officers send the statistical summary, narrative report, training records, and interception log for significant pests every month to their supervisors and RCPC, with copies to the following:

- Port director (monthly)
- NDDTC (only for 6 months after graduating from BCOT)
- Others designated by port policy

Local management must ensure that the information that is collected by the Canine Officer is timely and accurate. The Canine Officer must send a copy to the RCPC by the 10th of every month.

RCPCs use the operational reports to prepare annual reports for the regional directors with copies that go to the following:

- Regional Program managers
- NDDPM
- Port Directors
- SPHDs

Baggage Information Data (PPQ Form 277)

Purpose

Most work locations use PPQ Form 277, Baggage Information Data, to record daily activities performed by detector dog teams as a way of tallying the standard statistics for the monthly and end-of-year summaries and narrative reports. Refer to the standard statistics recorded for the *"Statistical Summary" on page-H-1-2.*

Basic information recorded is:

- Date
- Work site, such as airport, post office, and cargo
- Number of flights, passengers, bags, crew, and boxes screened
- Total passenger count from flight declaration
- Total number of responses—includes false and positive responses
- Number of positive responses
- Number of responses with seizures
- Number of quarantine material interceptions (QMIs)—plant, animal, and weight of the animal QMIs (in kilograms)
- Penalties and fines

Appendix H: Reporting and Documentation
Training Records

Canine officers should consult with supervisors, and/or RCPCs to find out the local procedures for tallying daily activities for detector dog teams. Otherwise, utilize the form to meet your needs. Some canine officers record the following additional information:

- Different responses, a new odor, or a residual odor
- Kind of response—crew, pit bag, false

For an explanation of PPQ Form 277, refer to the Airport and Maritime Operations Manual, Appendix 1.

Training Records

Purpose
To maintain records on detector dog training conducted by Canine Officers and National Canine Instructors. These records help identify the strengths and weaknesses of detector dogs.

Canine officers, as well as the national canine instructors, are responsible for recording the results of training sessions conducted with detector dogs. These training sessions may take place in a controlled environment, or even when training bags are positioned around a carousel during a flight. Refer to **Appendix K** for a full-size form you can copy.

Completing the Training Record Form
Figure H-1-2 illustrates the training record. The numbered items listed below refer to correspondingly labeled areas of the form.

Appendix H: Reporting and Documentation
Training Records

[Figure showing a blank Agriculture Detector Dog Training Record form labeled "EXAMPLE 1" with numbered keyed areas 1–9 indicating different sections of the form: header information (1), Exercise Type column (3), Con/Cnt/Pl column (4), Target Odors area (5), Total Trials (6), Non-Target Odors Response (7), Non-Target Odors area (8), and Remarks section (2,9).]

FIGURE H-1-2: Agriculture Detector Dog Training Record with Keyed Areas

1—List handler's name, date, canine's name, canine's weight, and port of assignment in the area provided.

2—Describe training objective for the session in the area labeled "REMARKS."

3—List type of exercise (general description of what you are doing, i.e. boxes: speed trial, mail: envelopes) in column headed "Exercise Type."

4—For each combination of Concentration, Container, and Placement fill in one row under the heading "Con/Cnt/Pl."

Consider placement when concentration and container have been addressed.

Container refers to texture/material of a case, i.e. Hard – a sturdy plastic luggage; Medium – a vinyl pit bag; and Soft – a cloth hand bag.

5—Note dog's response to each container with target items in the column "TARGET ODORS" under that target item.

Appendix H: Reporting and Documentation
Training Records

6—For each exercise record the number of times the team inspected a container with target items under the heading "Total Trials."

 If the dog did not sniff the container with target items it will not be considered inspected and not recorded as a trial but as a handler error.

7—List notable non-target odors included in exercise under the heading "NON-TARGET ODORS."

8—Record the dog's response to each non-target odor under the heading "NON-TARGET ODORS" – "Response."

9—Upon conclusion of a session detail any problems or record comments in the area labeled "REMARKS."

Example of a Completed Training Record

Figure H-1-3 is an example of a completed detector dog training record. An explanation of the entries follows the form.

EXAMPLE 2
AGRICULTURE DETECTOR DOG TRAINING RECORD

Handler	John Doe			Concentration	Container	Placement	Scoring			
Date	3/1/03			High	Hard	High	+ Positive Response		(+) Handler Assist	
Canine	Bombay			Medium	Medium	Low	- Non-Response		ı Handler Cue	
Weight	56 lbs			Low	Soft	Concealed	ı False Response		(-) Handler Error	
Port	San Juan PR			TARGET ODORS				NON-TARGET ODORS		
Exercise Type	Con/Cnt/Pl	Apple	Beef	Citrus	Mango	Pork			Response	Total Trials
(1) Luggage	H/H/L	++	++	-(+)	(+)+	++				10
(2) Luggage	M/H/L	++	++			++				10
	H/M/L			(+)+	++					
(3) Luggage										
Blank								Fish	ı	
(4) Luggage										
Extinction								Salmon		
								Tuna		
								Herring		

REMARKS
Objective
(1) Test proficiency on high concentration of five basic odors in hard sided suitcases
(2) Test citrus and mango, high concentration, medium containers and apple, beef, pork medium concentration, hard side suitcases
(3) Blank
(4) Extinction for fish

FIGURE H-1-3: Example of a Completed Agriculture Detector Dog Training Record

Appendix H: Reporting and Documentation
Training Records

(1) — The first exercise listed consisted of 50 suitcases. Five hard side suitcases were loaded with high concentrations of target items and placed low. Cheese and bread were included in this exercise as non-target items. Each suitcase was inspected by the dog two times.

- The team inspected suitcases containing target items a total of ten times. This number is recorded under the heading "Total Trials."
- The dog's response to each suitcase containing target items is listed under the target item heading. The dog gave a positive response each time he inspected a suitcase containing apple, beef, or pork. The dog gave no response the first time he inspected the suitcase containing citrus, and had to be assisted in order to give a positive response the second time he inspected that suitcase. The dog had to be assisted in order to give a positive response the first time he inspected the suitcase containing mango and gave a positive response the second time he inspected that suitcase.
- The dog did not respond to non-target odors.

The proficiency rate of the dog can be calculated by dividing the number of positive responses by the number of total trials. For this exercise the proficiency rate is 70%. Proficiency rates should only be calculated on exercises that contain target containers the dog is inspecting for the first time.

(2) — The second exercise listed consisted of 50 suitcases. Three hard side suitcases were loaded with medium concentrations of apple, beef, and pork and placed low. Two medium suitcases were loaded with high concentrations of citrus and mango and placed low. Chocolate was included in this exercise as a non-target item.

- The team inspected suitcases containing target items a total of ten times. This number is recorded under the heading "Total Trials."
- The dog's response to each suitcase containing target items is listed under the target item heading. The dog gave a positive response each time he inspected a suitcase containing apple, beef, mango, and pork. The dog had to be assisted to give a positive response the first time he inspected citrus and gave a positive response the second time he inspected that suitcase.
- The dog did not respond to non-target odors.

(3) — The third exercise listed consisted of 50 suitcases. There were no suitcases containing target items in the exercise. Cheese and fish were included in the exercise as non-target items.

- The team inspected each suitcase two times.
- The dog responded positively to a suitcase containing fish.

Appendix H: Reporting and Documentation
Health Care Records

(4) — The fourth exercise listed consisted of 50 suitcases. There were no suitcases containing target items in the exercise.

- The team inspected each suitcase two times.
- The dog did not respond to a suitcase containing fish.

Health Care Records

Health care records are those prepared and maintained by a veterinarian for detector dogs.

Purpose
The records provide the health and medical history of the detector dogs.

When and How to Maintain
Create a folder to hold the records. File them chronologically with the most recent date on top.

Distribution
Distribute the health care records for each detector dog as follows:

1. The original is maintained by the canine officer.
2. The RCPC reviews the health care records during port visits.
3. At the time of retirement, send a copy of the health care record to NDDTC.
4. NDDTC maintains a copy of the health care records on all detector dogs for 3 years after they are retired from service.
5. The original health care records are transferred with the dogs when they retire.

CBP Directive No. 3340-025B: Commissioner's Situation Room Reporting

Date: May 16, 2003

Supersedes: 3340-025A, 4/24/02

Review Date: May 2006

1: Purpose
To describe Customs and Border Protection (CBP) procedures requiring the timely reporting of incidents of terrorism, significant events, and emerging issues.

2: Authority/Background

The Commissioner established a "7x24" Situation Room at Headquarters which serves as the central CBP-wide point for reporting major occurrences to top management. (The authority regulating the issuance of this Directive is derived from the original memorandum dated October 30, 1998, FILE: MAN-1 OI: ICD KKQ, establishing the Situation Room.)

3: Scope/Definition

3.1 – The term "significant reporting" will apply to all information deemed necessary to advise the Commissioner and top management of incidents, significant events, and/or emerging issues.

4: Reporting Procedure

4.1 – Primary notifications will be made to the Commissioner's Situation Room prior to other established reporting entities.

4.2 – Keeping in mind the safety of our employees and the integrity of operations within the different disciplines of CBP, the following reporting procedures will be required.

4.3 – All CBP events related to terrorism must be reported telephonically and in writing immediately. All other incidents of a sensitive or timely nature will be reported telephonically to the Commissioner's Situation Room within 2 hours of its occurrence and followed by written notification within 4 hours. More serious events, which occur over an extended time period will require regular updates.

4.4 – The telephone number for the Commissioner's Situation Room is 1-877-748-7666 or 202-927-0425. Reports are to be faxed directly to the Commissioner's Situation Room at 202-927-5477. For classified reports, fax directly to 202-927-2692.

5: Written Response

Attached is an updated template detailing the information and narrative expected in the written notification.

6: Responsibilities

6.1 – The Assistant Commissioners for Border Patrol and Field Operations, through their respective executive staffs, are responsible for ensuring that the Commissioner's Situation Room is notified in a timely manner. The specific responsibility to telephonically notify and provide written follow-up to the Commissioner's Situation Room rests with the first-line supervisor. This responsibility may be delegated only on occasions when the first-line supervisor is not present.

Appendix H: Reporting and Documentation
CBP Directive No. 3340-025B: Commissioner's Situation Room Reporting

This requirement to notify the Commissioner's Situation Room does not obviate the need to report such events through the normal chain of command. Simultaneous submission of written reports to the Assistant Commissioner for Border Patrol, the Assistant Commissioner for Field Operations, and others through the chain of command is expected and is operationally appropriate.

6.2 – The Assistant Commissioner for Internal Affairs is the primary management official designated to receive and advise the Commissioner of reports relating to the physical security of CBP facilities, or integrity issues involving CBP personnel. Significant reports dealing with these issues will continue to be transmitted to the Assistant Commissioner for Internal Affairs through the Office of Internal Affairs Significant Activity Reporting system.

7: Reportable Items

7.1 – While it is difficult to provide an all-inclusive list of the types of incidents, events, or issues that should be reported, the following types of significant incidents should be reported:

7.1.1 – any terrorism-related incident, including, but not limited to the following;

7.1.1.1 – any seizure of, or any situation or incident or other enforcement action associated with, a potential Weapon of Mass Destruction (WMD), including a chemical, biological, radiological, nuclear or explosive (CBRNE) device, or a precursor or component of such a device;

7.1.1.2 – any seizure of funds connected to suspected terrorists or terrorist organizations;

7.1.1.3 – any arrest or detention of a terrorist, or suspected terrorist;

7.1.1.4 – any seizure based upon antiterrorism enforcement initiatives, regardless of value or quantity;

7.1.1.5 – any seizure, arrest, or detention based upon a CBP antiterrorism targeting effort;

7.1.1.6 – any seizure, arrest, or detention resulting from standard CBP enforcement activities that are, in the judgment of the reporting supervisor, potentially related to terrorism;

7.1.1.7 – any passenger, cargo, and conveyance examination in which potential terrorism-related documents or material is obtained, regardless of whether the examination resulted in a seizure or arrest for a CBP violation;

7.1.1.8 – any specific intelligence received indicating that a suspected terrorist, WMD or precursor component, or explosive device, will enter or depart the U.S. by airplane, vessel, vehicle, or pedestrian traffic at a specific time or place—or that any dangerous device has been placed at a port of entry;

7.1.1.9 – any terrorist threat received by CBP personnel, including any threat against CBP facilities, equipment, personnel, or operations;

7.1.1.10 – any significant request for assistance and operational response from another agency that is antiterrorism related;

7.1.1.11 – any anticipated or ongoing situation related to potential terrorist activity that may involve or require significant operational coordination with another agency or with foreign authorities;

7.1.1.12 – any anticipated or ongoing situation related to potential terrorist activity that may involve significant scrutiny or attention from the media, the trade, or the public;

7.1.1.13 – any discovery of a suspected WMD in a Federal Inspection Services area, a cargo inspection, or mail facility;

7.1.1.14 – any incident or activity not specifically addressed that in the judgment of the reporting supervisor has the potential to contribute to the interagency effort to combat terrorism;

7.1.2 – any radiation detection incident where the Office of Information and Technology Labs and Scientific Services Division has determined that a potential nuclear or radiological alarm warrants a request for response from the Department of Energy, as outlined in the National Radiation Detection Program standard operating procedures;

7.1.3 – any death of or major injury to a CBP employee on or off duty;

7.1.4 – any shooting incidents involving CBP employees, to include accidental discharges;

7.1.5 – the death, injury, or escape of an individual which was caused by the actions of CBP personnel (either on or off duty) or occurred while the individual was detained in CBP custody (including suicide attempts);

7.1.6 – any assault of a CBP employee occurring in relation to his/her employment or official duties;

7.1.7 – the arrest, detention or incarceration of a CBP employee by authorities;

7.1.8 – **any lost or missing CBP Detector Dogs;**

7.1.9 – **any death of or major injury to CBP Detector Dog on or off duty;**

7.1.10 – **the quarantine or detention of a CBP Detector Dog;**

7.1.11 – **any bite or injury caused by a CBP Detector Dog;**

7.1.12 – any positive identification of animal or plant diseases that may have serious agricultural and economic consequences in the United States and for which immediate action and notification is required;

7.1.13 – any declared airborne/marine emergency or incident resulting in property damage;

7.1.14 – any unscheduled port/office closing or significant disruptions to port operations for reasons to include but not limited to, bomb threats, public demonstrations, systems failures, weather and environmental hazards;

7.1.15 – unscheduled disruption of services to trade due to Automated Cargo System brownouts or outages, which exceed 2 hours. Major disruption of other automated systems on a national or regional basis as determined and reported by the Office of Information and Technology;

7.1.16 – any unscheduled system shutdowns of a fixed radiation detection device, (e.g., a radiation portal monitor) greater than 24 hours;

7.1.17 – the seizure of a foreign or domestic commercial conveyance;

7.1.18 – seizures of more than

- 500 kilograms of marijuana
- 50 kilograms of cocaine
- 50 kilograms of methamphetamine/amphetamine
- 200 kilograms of hashish
- 500 kilograms of khat
- 2 kilograms of heroin
- 2 kilograms of opium
- 2 kilogram of MDMA (ecstasy)
- 1 million dosage units of other dangerous drugs

- $250,000 in currency or negotiable instruments
- $500,000 in real property or merchandise
- $1 million penalty
- stolen cars outbound (value in excess of $250,000);

7.1.19 – any foreign military or law enforcement incursions;

7.1.20 – any crewmember who deserted or after being detained onboard, absconded;

7.1.21 – any significant seizure of child pornography worthy of media attention;

7.1.22 – any other event that may warrant review by senior management to include but not limited to heroic acts and/or public recognition (e.g. rescues, significant results of joint operations);

7.1.23 – any event or incident which may be politically sensitive to the U.S. or foreign government to include searches and detentions of persons claiming diplomatic immunity or special status, requests for asylum made to CBP officials; actions involving foreign or U.S. government officials, government representatives, or prominent foreign nationals; and

7.1.24 – any event or incident that has resulted or may result in significant media attention.

8: Controlled Deliveries
The CBP supervisor is responsible to report seizures which meet the aforementioned criteria and should advise whether a controlled delivery is pending in the synopsis of incident (see attached report form.)

9: Requests for Information
One indicator of potential emerging issues may be Freedom of Information Act requests or other inquires, particularly those received from Congress, the press, or various advocacy groups. Requests of this type which seem to be indicative of a potential emerging management issue or problem should be reported.

Commissioner of Customs and Border Protection

Attachments

Appendix H: Reporting and Documentation
CBP Directive No. 3340-025B: Commissioner's Situation Room Reporting

Instructions for Completing Significant Incident Report

1. Provide the date, time and exact location of the significant incident. Provide the date, time and name of person in the Commissioner's Situation Room to whom the incident was reported telephonically. When telephonically reporting significant incidents to the Commissioner's Situation Room, you must obtain a Commissioner's Situation Room reference number. Place this number in the top right hand box on the report form.

2. Provide the Designated Field Office or Sector and Port of Entry or Station that is submitting the report. Provide the name and telephone number of the person submitting the report. Provide a point of contact and telephone number in the event additional information not contained in the report is needed.

3. Check the appropriate box as to the type of incident and if it involved CBP personnel, indicate whether it happened on or off duty.

4. Provide a brief synopsis of the incident to include the names of individuals involved, commodity, weight, and value (if known) of items seized. If an arrest is made, indicate the number, sex, and citizenship of those arrested. It is important that the questions; who, what, where, when, why are answered.

5. Indicate notifications made either telephonically, by fax or E-mail (i.e. telephonic to Commissioner's Situation Room, Office of Field Operations by fax). Provide the date, time, and telephone number.

6. List the names of fatalities or those injured as they relate to #4 above.

7. Indicate what action has been taken as a result of the incident (i.e., in the case of a narcotic seizure, "turned over to ICE for controlled delivery"). In the event of serious injury of employees, indicate the name and phone number of the hospital involved.

Appendix H: Reporting and Documentation
CBP Directive No. 3340-025B: Commissioner's Situation Room Reporting

U.S. DEPARTMENT OF HOMELAND SECURITY
Bureau of Customs and Border Protection

SIGNIFICANT INCIDENT REPORT
CBP Directive 3340-025B

1. DATE OF INCIDENT: _____ LOCATION OF INCIDENT: _____ CSR NUMBER: _____
TIME OF INCIDENT: _____
REPORTED TO COMMISSIONER'S SITUATION ROOM VIA PHONE ON:
DATE: _____ TIME: _____ TO: _____

2. REPORTING OFFICE: _____ DFO/SECTOR: _____ POE/STATION: _____
PERSON MAKING REPORT: _____
OFFICE PHONE: _____ CELL PHONE: _____ FAX NUMBER: _____
POINT OF CONTACT: _____
OFFICE PHONE: _____ CELL PHONE: _____ FAX NUMBER: _____

3. TYPE OF INCIDENT: ☐ ON DUTY ☐ OFF DUTY
☐ TERRORIST RELATED   ☐ SHOTS FIRED AT OR BY EMPLOYEE   ☐ SIGNIFICANT AGRICULTURAL EVENT   ☐ SUICIDE ATTEMPT
☐ EMPLOYEE ARRESTED   ☐ CANINE INCIDENT   ☐ CONVEYANCE/AIRCRAFT INCIDENT   ☐ MEDIA INTEREST
☐ EMPLOYEE ASSAULTED   ☐ SIGNIFICANT SEIZURE   ☐ FOREIGN MILITARY/POLICE INCURSION ☐ CREW DESERTERS
☐ EMPLOYEE DEATH   ☐ SIGNIFICANT ARREST/DETENTION   ☐ BOMB THREAT   ☐ ESCAPE
☐ EMPLOYEE INJURED   ☐ NON-EMPLOYEE INJURY/DEATH   ☐ FACILITY DISRUPTIONS   ☐ OTHER:_____
☐ RADIATION DETECTION EVENT ☐ RESCUE   ☐ TECHNOLOGY DISRUPTIONS

4. SYNOPSIS: (USE CONTINUATION SHEET IF NECESSARY) _____

SEIZURE TYPE: _____ QUANTITY: _____ VALUE: _____
NUMBER OF ARRESTS: _____ MALE: _____ FEMALE: _____ CITIZENSHIP: _____

5. NOTIFICATIONS MADE:
1. ☐ **_TELEPHONIC REPORT TO COMMISSIONER'S SITUATION ROOM   (202) 927-0425_**
2. ☐
3. ☐
4. ☐
5. ☐
6. ☐
7. ☐
8. ☐

SAMPLE

6. INJURIES/FATALITIES:
NAME AND EXTENT OF INJURY:
1.
2.
3.
NAME OF FATALITIES:
1.
2.
3.

7. ACTION TAKEN _____

CBP Form 6 (05/03)

FIGURE H-1-4: Significant Incident Report

Appendix H: Reporting and Documentation
CBP Directive No. 3340-025B: Commissioner's Situation Room Reporting

---

SIGNIFICANT INCIDENT REPORT
Continuation Sheet

PAGE _____ OF _____

DATE OF INCIDENT:	LOCATION OF INCIDENT:	CSR NUMBER:
TIME OF INCIDENT:		

SYNOPSIS: (CONTINUED)

CBP FORM 6 (CONT)(05/03)

FIGURE H-1-5: Significant Incident Report Continuation Sheet

Appendix I

Manual Maintenance

Contents

Introduction **page-I-1-1**
Issuing Revisions **page-I-1-1**
Keeping Manuals Current **page-I-1-1**
Knowing What's Revised **page-I-1-2**
Knowing Your Responsibility **page-I-1-2**
Ordering Manuals **page-I-1-3**
Adding and Changing Addresses and Copy Counts **page-I-1-3**
Correcting Errors and Suggesting Improvements **page-I-1-4**

Introduction

This appendix describes how APHIS-PPQ will support this manual. Directions for you to follow in maintaining the integrity of the National Detector Dog Manual are included in this appendix.

Issuing Revisions

APHIS-PPQ will revise the National Detector Dog Manual by distributing immediate updates received from the NDDPM. We will schedule new editions at fixed intervals—at least every 5 years. If more than 50 percent of a section changes, we will issue a new section. We will **not** issue an update solely to correct a minor typographical error. Errors will be corrected only when they would lead to an incorrect action.

Keeping Manuals Current

There are three ways to track revisions to this manual—the *Update Record*, transmittal memos, and control data.

The **Update Record** is on the back of the title page. Use it to record all the transmittals you receive. If you miss a transmittal, the *Update Record* alerts you.

APHIS-PPQ will mail all revisions with a transmittal memo. The memos are numbered consecutively—allowing you to know if you have missed a transmittal. Filing these memos to assure that you have received all the previous issuances is best. File transmittals immediately upon receipt.

Appendix I: Manual Maintenance
Knowing What's Revised

Besides having numbered transmittals, each page in the manual has control data. This is positioned at the bottom of the page. The control data on revised pages alerts you to whether you have the most up-to-date version. Control data look like this:

02/2003-01
PPQ

02/2003 is the month and year the page was issued. -01 is the transmittal number. The first transmittal issued for the year is always (month and year)-01.

Knowing What's Revised

The transmittal will explain the revision's purpose and give you directions for making the revision.

Except changes to the index, APHIS-PPQ marks all revisions with change bars next to the altered text in the left margin. If no other changes occur, material moved from the bottom of one page to the top of the next page will not be marked.

Knowing Your Responsibility

To enhance professionalism, keep your National Detector Dog Manual current. Therefore, please do the following:

1. Read the revisions when you receive them.
2. Record your transmittal in the **Update Record**.
3. Add or replace the revised pages the day you receive them.
4. If a practice exercise is included, complete it.
5. File transmittal memos in your manual.
6. If you miss a transmittal, order another one.
7. Let your RCPC know when PPQ's Manuals Unit has made an error. The RCPC will notify the RPM, who will notify the NDDPM.
8. Give your RCPC your suggestions for improvements. The RCPC will notify the RPM, who will notify the NDDPM.

Ordering Manuals

The NDDPM, in partnership with PPQ's Manuals Unit, is responsible for maintaining and distributing the National Detector Dog Manual.

The address of the NDDPM is as follows:

USDA, APHIS, PPQ, Port Operations
4700 River Road Unit 60
Riverdale, MD 20737-1236
Ph: 301-734-8295
FAX: 301-734-5269

Contact PPQ's Manual Unit:

USDA, APHIS, PPQ
69 Thomas Johnson Drive, Suite 100
Frederick, MD 27102-4301
Ph: 240-629-1929
FAX: 301-663-3240

Use E-mail, FAX, telephone, or mail when requesting services, and always provide the following:

Organization
P.O. Box or Street Address, include Room or Suite Number
City, State, and nine-digit Zip Code
Contact Person
Telephone Number
FAX Number

When ordering the manual and related updates (transmittals), provide the following additional information:

- List the title: National Detector Dog Manual
- Indicate either the initial manual or a transmittal number
- List the number of copies you need

Adding and Changing Addresses and Copy Counts

When adding and changing addresses and copy counts for distribution, provide the following additional information:

- List the title: National Detector Dog Manual
- List the number of copies you need to get
- List the new, corrected, or deleted address

Correcting Errors and Suggesting Improvements

If you detect an error, report it by using a Comment Sheet that is included with this manual. Or, if it is easier, call, send an E-mail message, or FAX the PPQ's Manuals Unit with cc to the NDDPM.

Do the same if you want to suggest an improvement or question a procedural change. If your improvement is substantive, you might want to submit a formal suggestion to your RCPC, who will forward to the RPM and then to the NDDPM.

Appendix J

Legislative Authority

Contents

Barney Bill   page-J-1-1
Animal Welfare Act (AWA)   page-J-1-2

Barney Bill

The following is an excerpt from Bill H. R. 2559 (commonly known as the Barney Bill).

Title V–Inspection Animals

Sec. 501. Civil Penalty

(a) In General. Any person that causes harm to, or interferes with, an animal used for the purposes of official inspections by the Department of Agriculture, may, after notice and opportunity for a hearing on the record, be assessed a civil penalty by the Secretary of Agriculture not to exceed $10,000.

(b) Factors in Determining Civil Penalty. In determining the amount of a civil penalty, the Secretary shall take into account the nature, circumstance, extent, and gravity of the offense.

(c) Settlement of Civil Penalties. The Secretary may compromise, modify, or remit, with or without conditions, any civil penalty that may be assessed under this section.

(d) Finality of Orders.

(1) In general. The order of the Secretary assessing a civil penalty shall be treated as a final order reviewable under chapter 158 of title 28, United States Code. The validity of the order of the Secretary may not be reviewed in an action to collect the civil penalty.

(2) Interest. Any civil penalty not paid in full when due under an order assessing the civil penalty shall thereafter accrue interest until paid at the rate of interest applicable to civil judgments of the courts of the United States.

Appendix J: Legislative Authority
Animal Welfare Act (AWA)

Sec. 502. Subpoena Authority

(a) In General. The Secretary shall have power to subpoena the attendance and testimony of any witness, and the production of all documentary evidence relating to the enforcement of section 501 or any matter under investigation in connection with this title.

(b) Location of Production. The attendance of any witness and the production of documentary evidence may be required from any place in the United States at any designated place of hearing.

(c) Enforcement of Subpoena. In the case of disobedience to a subpoena by any person, the Secretary may request the Attorney General to invoke the aid of any court of the United States within the jurisdiction in which the investigation is conducted, or where the person resides, is found, transacts business, is licensed to do business, or is incorporated, in requiring the attendance and testimony of any witness and the production of documentary evidence. In case of a refusal to obey a subpoena issued to any person, a court may order the person to appear before the Secretary and give evidence concerning the matter in question or to produce documentary evidence. Any failure to obey the court's order may be punished by the court as a contempt of the court.

(d) Compensation. Witnesses summoned by the Secretary shall be paid the same fees and mileage that are paid to witnesses in courts of the United States, and witnesses whose depositions are taken, and the persons taking the depositions shall be entitled to the same fees that are paid for similar services in the courts of the United States.

(e) Procedures. The Secretary shall publish procedures for the issuance of subpoenas under this section. Such procedures shall include a requirement that subpoenas be reviewed for legal sufficiency and signed by the Secretary. If the authority to sign a subpoena is delegated, the agency receiving the delegation shall seek review for legal sufficiency outside that agency.

(f) Scope of Subpoena. Subpoenas for witnesses to attend court in any judicial district or testify or produce evidence at an administrative hearing in any judicial district in any action or proceeding arising under section 501 may run to any other judicial district.

Animal Welfare Act (AWA)

You can find the AWA at the following website:

http://www.aphis.usda.gov/ac/awapdf.pdf

Appendix K

Forms

Introduction

This appendix contains blank forms that you can copy as needed. Remember to replace the forms in your binder for future reference.

Telephone Interview Worksheet

Section 1: Instructions

When someone calls wishing to donate a dog to the USDA Detector Dog Program, use this Telephone Interview Worksheet to assess the dog's general traits. Photocopy this worksheet and use it to record answers and take notes.

Beginning
1. Confirm that you are speaking to the owner or primary caregiver of the dog, since that person will know the most about the dog.
2. Let the owner know that you have a series of questions to ask that will help you learn a little more about the dog. You will need about 20 minutes of the owner's time.

During
1. Let the owner offer information about the dog; sometimes the owner will answer questions not yet asked.
2. Listen carefully while keeping the interview on track.
3. Do not prompt the answers; let the owner answer the questions.
4. Take notes during the telephone interview, especially if there is a time gap between the interview and the appointment to meet the dog.

End
Use the following table to determine how to end the telephone interview.

If the interview was:	Then:
Successful	Schedule an appointment to meet the dog in its home environment
	(This meeting will provide the evaluator a baseline of the dog's character. A dog may act bold and courageous in its own environment, but become fearful in a public place such as an airport.)
Unsuccessful	Thank the owner and explain why the dog is unacceptable

Section 2: Questions

1. What is the dog's name? _____

HINT: Once you know the dog's name, use it throughout the interview. Remember that the caller is considering donating a member of the family.

---

2. Is **[Dog's Name]**:

    A. Male ❒ Female ❒

    B. Spayed or Neutered? Yes ❒ No ❒

3. How old is **[Dog's Name]**? _____ Years, and/or _____ Months

 The dog **MUST** be between 9 months and 3 years old.

If the dog is outside of this range, **STOP THE INTERVIEW**. Thank the owner and explain why the dog is unacceptable.

4. Where did you get **[Dog's Name]**? _____

   A. At what age? _____

HINT: Knowing the dog's age when it was originally acquired will help determine the timing, quality, and quantity of its exposure to socializing factors.

5. Did you ever take **[Dog's Name]** on outings to a park or to obedience school? _____

   A. Yes, go to 5 C, then 5 D.

   B. No, go to 6.

   C. How old was **[Dog's Name]** when it went on these types of outings? _____

   D. How did **[Dog's Name]** behave during these types of outings?
   _____
   _____
   _____

6. How often did you take **[Dog's Name]** to the veterinarian? _____

HINT: The answer will help determine if the dog was regularly vaccinated. Also, if the dog has only been exposed to veterinarian visits, it may not have good social skills.

7. Have you ever seen **[Dog's Name]** have a seizure, or are you aware of any history of seizures? _____

If the dog has had:	And there is:	Then:
A seizure	➝	**1. STOP THE INTERVIEW** 2. Thank the owner and explain why the dog is unacceptable
No seizure	A history of seizures	
	No history of seizures	Continue to 8

8. Do you give **[Dog's Name]** heartworm preventive medicine year round? _____

If the dog:	Then:
Has been on preventive heartworm medicine year round	Continue to 9
Has **not** been on preventive heartworm medicine	**1. STOP THE INTERVIEW** 2. Ask the owner to take the dog to a veterinarian to have an occult heartworm test and to provide you with the test results before continuing the procurement process **NOTE**: If there is financial hardship, this test can be conducted at the expense of the USDA, provided the rest of the interview is positive

9. To the best of your knowledge, does **[Dog's Name]** have any health problem? _____

If the dog:	Then:
Has no health problem	Continue to 10
Has a health problem	1. Ask the owner to explain the health problem 2. Ask if you could speak directly to the attending veterinarian 3. If the owner agrees, have the owner call the veterinarian in advance to give the doctor permission to speak with you 4. **STOP THE INTERVIEW** until you can consult with the veterinarian and the NDDTC

10. Do you allow **[Dog's Name]** to interact with guests at your home? _____

    A. If yes, how does the dog react? _____

    B. If no, why? _____

If the dog was:	And the dog's reaction was:	Because the dog is too:	Then:
Allowed to interact with guests	Pleasant, bold, or obnoxious	→	Continue to 11
	Frightened and/or submissive, urinates, or tucks its tail	→	**1. STOP THE INTERVIEW** 2. Thank the owner and explain why the dog is unacceptable
Not allowed to interact with guests	→	Bold or obnoxious towards guests	Continue to 11
		Frightened and/or shy, demonstrating behavior such as submissive urinating	**1. STOP THE INTERVIEW** 2. Thank the owner and explain why the dog is unacceptable

11. Do you have children or does **[Dog's Name]** interact with children often (at least once a week)? _____

    A. If yes, how does the dog react? _____

    B. If no, why? _____

> **CAUTION** If the answer to this or any other question of the interview sets off an alarm, note the alarm on this worksheet and consider it when initially screening the dog.

Note alarms and considerations for initial screening:
_____
_____
_____

If the children are:	And the dog demonstrated:	Then:
Twelve years old or younger	Fear or aggression	1. Note an alarm 2. Consider that this age group can (a) be inexperienced with dogs, and (b) have voices that are squeaky and tend to illicit play bite tendencies 3. Continue to 12
	Little to no fear or aggression	1. No alarm 2. Continue to 12
Teenagers	Fear or aggression	1. Note an alarm 2. Consider that (a) friends of teens can be interpreted as a stranger to a dog, and (b) sometimes teens observe fads that may cause a dog to exhibit a protective defense behavior in the home environment 3. Continue to 12
	Little to no fear or aggression	1. No alarm 2. Continue to 12
Young adults	→	1. No alarm 2. Consider as adults, not children 3. Continue to 12

12. What type of food do you feed **[Dog's Name]**? When you feed **[Dog's Name]**, does it eagerly gobble up the food or does it pick at the food?
_____
_____

HINT: You are looking for a dog that gobbles its food until it is gone. A dog that picks at its food, although not a good sign, may do so for several reasons. Therefore, use the following table for all the variables to consider.

> **CAUTION** If the answer to this or any other question of the interview sets off an alarm, note the alarm on this worksheet and consider it when initially screening the dog.

Note alarms and consideration for initial screening:
_____
_____

If the type of food is:	And the dog:	And the dog:	Then note that:
Dry	Gobbles the food	→	1. The dog has a strong food drive 2. Continue to 13
	Picks at the food	Eats around another animal that is dominant	1. The dog's food drive may be stifled 2. Note an alarm 3. Continue to 13
		Does not eat around another animal that is dominant	1. The dog is either over weight or has a weak food drive 2. Note an alarm 3. Continue to 13
Wet	Gobbles the food	→	1. The dog may not have a true, strong food drive 2. Note an alarm 3. Continue to 13
	Picks at the food	Eats around another animal that is dominant	1. The dog's food drive may be stifled 2. Note an alarm 3. Continue to 13
		Does not eat around another animal that is dominant	1. The dog is either over weight or has a weak food drive 2. Note an alarm 3. Continue to 13
Both dry and wet	Gobbles the food	→	1. The dog may not have a true, strong food drive 2. Note an alarm 3. Continue to 13
	Picks at the food	Eats around another animal that is dominant	1. The dog's food drive may be stifled 2. Note an alarm 3. Continue to 13
		Does not eat around another animal that is dominant	1. The dog is either over weight or has a weak food drive 2. Note an alarm 3. Continue to 13

13. Why are you considering donating **[Dog's Name]** to the USDA?
_____
_____
_____
_____

HINT: This is a good, general question to end the interview. The answer may be enlightening.

United States Department of Agriculture

Marketing and Regulatory Programs

Animal and Plant Health Inspection Service

Plant Protection and Quarantine

Subject: Limited Release Form and Sterilization Agreement

I _____ do hereby give permission to
     (Owner)

_____ of the U.S. Department of Agriculture
   (USDA Representative)

to take _____, _____
     (Name of Dog)   (Breed of Dog)

off my property for the sole purpose of temperament testing and health screening. Health screening will be done at no cost to me, the dog's owner. It is my understanding that I am **not** relinquishing legal claim or ownership at this time. I understand that if the dog is **not** accepted into the USDA Detector Dog Program, the dog will be returned to me at my expense. However, if the dog is accepted into the Program, the USDA assumes the responsibility to have

_____ spayed or neutered. At that time, a final release
  (Name of Dog)

statement will be signed relinquishing my legal claim to the dog.

_____   _____
USDA Representative   Owner/Agent

_____
PPQ Work Location

Date: Date:

Initial Screening Process Worksheet

> **CAUTION** The evaluator should conduct this initial screening with either the owner or shelter staff available in the event the dog becomes aggressive. If it does, **STOP THE EVALUATION**. Do **not** attempt to continue.

Section 1: Instructions
Complete the following information:

Dog's Name: _____
Dog's Sex: _____
Evaluation Date: _____
Evaluation Location: _____
Evaluator's Name: _____

Purpose
Determine if the dog initially meets the criteria as a potential candidate for the Agency's Detector Dog Program. The criteria are high food drive, sociability, ability to train, physical soundness, and low anxiety level.

When to Conduct
Conduct the initial screening after completing the Telephone Interview Worksheet and before the temperament evaluation is conducted at an airport.

Completed Each Evaluation Step
In Sections 2, 3, 4, and 6: choose either FAIL, AVERAGE, or EXCELLENT. In section 5: choose either NO or YES

> **Important** If the dog fails any part of this initial screening, **STOP THE EVALUATION**.
>
> Arrange for the return of the dog to its owner at the owner's expense.

Section 2: High Food Drive
Goal: Since the reward (motivation) for detector dogs is food, food must be the dog's priority in life.

1. Feed dog treats (use a wide variety of treats). The dog:

Spits treats out; not interested	Fail	❏
Consistently takes treats	Average	❏
Gobbles treats; eagerly anticipates next treat	Excellent	❏

2. Show dog the treats; then:
    A. Place treats up high
    B. Place treats low
    C. Hide treats underneath something. The dog:

Does not search for treats	Fail	❏
Searches with encouragement	Average	❏
Jumps, digs for treats	Excellent	❏

3. Feed dog treats and have a stranger (i.e., person unknown to dog) distract the dog (do **not** call dog by name). The dog:

Is distracted by stranger; ignores food	Fail ❏
Hesitates between stranger and food	Average ❏
Chooses food; ignores stranger	Excellent ❏

Section 3: Sociability

Goal: Detector dogs must demonstrate self-confidence around all types of people (i.e., different ages, races, sexes, persons with disabilities).

1. Greet the dog (initial greeting). The dog:

Does not approach; displays submissive urinating	Fail ❏
Approaches hesitantly, but recovers upon interaction (3-5 seconds)	Average ❏
Exhibits obvious friendly demeanor to all types of people	Excellent ❏

2. Observe dog's reaction to environment (people, vehicles, noises). The dog:

Is afraid; doesn't immediately recover (3-5 seconds)	Fail ❏
Rarely startles; recovers immediately	Average ❏
Remains stable; comfortable in environment	Excellent ❏

3. Observe dog's reaction to stranger in the following situation:

   > **CAUTION** Prepare to protect yourself. The following test is designed to determine a dog's aggressive or submissive tendencies.

   A. Attach dog, with a 4 to 6 foot leash, to a fence
   B. Have stranger stand 20 to 30 feet from dog
   C. Tell stranger to make eye contact with dog while maintaining distance
   D. Have stranger act in an unusual manner (make loud noises, move from side-to-side, wave arms) and advance toward dog, but never closer than 10 feet from dog
   E. Tell stranger to stop acting in an unusual manner
   F. Indicate to the stranger to now act in a friendly manner (change expression in face, change tone in voice, and discontinue eye contact), move toward the dog while maintaining a 10 foot distance
   G. If it is obvious that the dog poses no threat, the stranger can approach and pet the dog (optional)

Retreats and/or shows any aggression toward stranger; urinates submissively	Fail ❏
Startles; backs up a few steps during stranger's unusual manner, but when stranger acts friendly immediately wants to greet stranger	Average ❏
Maintains friendly posture; does not startle	Excellent ❏

4. Determine negative conditioning:
    A. Raise your hand back
    B. Quickly move your hand towards the dog's face, but do **not** actually strike dog. The dog:

Cowers and goes to ground, and/or displays submissive urinating	Fail	❏
Blinks and/or squints, but maintains friendly posture	Average	❏
Does **not** blink or squint; shows no signs of abuse	Excellent	❏

Section 4: Intelligence

Goal: The dog must be able to comprehend and complete repetitive tasks. Conduct two tests to determine the dog's aptitude for scent work.

1. Ask permission to release the dog in an enclosed/fenced area if the tests are being conducted at a shelter.
2. Take the dog to the area, preferably secured with some high grass.
3. Allow dog to relieve itself.

Test 1:

4. Show dog treats.
5. Put treats on ground (space around).
6. Unleash dog (prefer unleashing dog; but dog can remain leashed).
7. Tell dog to "Find It." Repeat the command 3 times.

Test 2:

8. Leash dog.
9. Cover dog's eyes with your hand or turn dog away from the direction where you will throw treats.
10. Throw treats in a random fashion into the wind so they land in high grass.
11. Unleash dog and observe tracking techniques.

Could not/did not find treats; more interested in other things	Fail	❏
Located some treats	Average	❏
Located most or all treats in short period of time; diligent in finding all treats	Excellent	❏

Section 5: Physical Soundness

1. Examine appearance of dog.
    A. Look for overall symmetry.
    B. Stand 5 feet from dog.
    C. Look at dog from side to side.
    D. Look at dog from front to rear.

Dog is well balanced	No	❏
	Yes	❏
Dog's front is in proportion to rear	No	❏
	Yes	❏

2. Examine dog's nails to determine general condition, with the exception of the dewclaws (abnormal wear on left or right side of either set of paws can denote a compensation for an abnormality).

   A. Are some nails different lengths than other? If yes, go to D.   No ❏
   Yes ❏

   B. Are the nails on one front paw the same length as those on the other front paw? If no, go to D.   No ❏
   Yes ❏

   C. Are the nails on one rear paw the same length as those on the other rear paw? If no, go to D.   No ❏
   Yes ❏

   D. Indicate which paw(s) has abnormal wear on the nails.

   Left front paw ❏   Right front paw ❏
   Left rear paw ❏   Right rear paw ❏

3. Examine dog's teeth to determine an approximate age.

   A. Do teeth show excessive wear (i.e., very rounded tips, bottom incisors have observable pulps)?   No ❏
   Yes ❏

   B. Do gums appear to be receding?   No ❏
   Yes ❏

If your answer to any of the above questions was:	Then:
Yes	1. **STOP THE EVALUATION** 2. Have veterinarian examine dog to verify dog's age during the health evaluation 3. If the dog is over 3 years old, it can **not** be accepted into program 4. Arrange for the return of the dog to its owner at the owner's expense
No	1. Do not contact veterinarian at this time 2. Continue initial screening

4. Look at dog's eyes for:

   A. Excessive tearing (watering) of one or both eyes   No ❏
   Yes ❏

   B. Tumors on one or both eyes   No ❏
   Yes ❏

   C. Cataracts on eyes   No ❏
   Yes ❏

   D. Entropion (eyelids curve inward) on one or both eyes   No ❏
   Yes ❏

   E. Ectropion (eyelids are droopy) on one or both eyes   No ❏
   Yes ❏

If your answer to any of the above questions was:	Then:
Yes	1. **STOP THE EVALUATION**
	2. Consult with the NDDTC for assistance
No	Continue initial screening

Section 6: Anxiety Level
Goal: Dog must be content in a crate or kennel.

1. Observe dog's level of anxiety while in crate or kennel.
    A. Place dog in crate or kennel.
    B. Offer dog treats.
    C. Leave the room for 3 to 5 minutes.
    D. Return to the room and offer dog treats again. The dog:

    Exhibits extreme signs of stress, such as biting and salivating;
    will not take treats in crate/kennel, even if person is in room — Fail ❏

    Whines, cries, but settles down; eats treat when offered — Average ❏

    Settles down immediately, comfortable in crate/kennel;
    takes treat when offered (signs that dog has previously been
    crate trained) — Excellent ❏

Contact the RCPC and the NDDTC for instructions to proceed.

Temperament Evaluation Worksheet

Important

The temperament evaluation should be conducted at an airport by an RCPC or a designated Canine Officer.

Evaluator's Name:_____   Evaluation Date:_____

Dog's Name: _____   Evaluation Location:_____

Alias: _____   Age: _____

Weight: _____   Sex: _____

Recommended Weight: _____   Breed: _____

Section 1: Reactions to Various Stimuli/Situations

Section 1 is divided into five parts (A-E). Rate the dog's reaction to each of the listed stimuli or situations under each part. Use a scale of 1 to 5 where the rating of 1 means poor and 5 means excellent. **Circle** the number that represents your rating. At the end of each part, **add** your ratings, **divide** the sum by the total number of items in that part to get a mean (average) rating, and **record** the mean rating in the space provided. If the sum of the mean ratings is 18 or above, then fax a copy of this worksheet the NDDTC and receive guidance whether to continue on to evaluating the dog's health.

Part A: Food Incentive

Food Incentive Stimuli or Situations	Poor	Fair	Average	Good	Excellent
Takes food from hand	1	2	3	4	5
Takes food from floor	1	2	3	4	5
Takes food from under baggage	1	2	3	4	5
Takes food up high	1	2	3	4	5
Takes food while on conveyor belt	1	2	3	4	5
Takes food on/around carousel	1	2	3	4	5
Takes food under stress	1	2	3	4	5
Other	1	2	3	4	5

Food Incentive Mean Rating: (Must achieve a mean rating of 4 or above)

Part B: Social

Social Stimuli or Situations	Poor	Fair	Average	Good	Excellent
Children	1	2	3	4	5
Adults	1	2	3	4	5
Small groups	1	2	3	4	5
One-on-one	1	2	3	4	5
Playfulness	1	2	3	4	5
Willingness to follow	1	2	3	4	5
Other	1	2	3	4	5

Social Reaction Mean Rating: (Must achieve a mean rating of 4 or above)

Part C: Environmental

Environmental Stimuli or Situations	Poor	Fair	Average	Good	Excellent
Baggage carts	1	2	3	4	5
Baggage tugs	1	2	3	4	5
Baggage carousels	1	2	3	4	5
Doorways	1	2	3	4	5
Tight quarters	1	2	3	4	5
Strange/new areas	1	2	3	4	5
Verbal praise	1	2	3	4	5
Tactile stimuli	1	2	3	4	5
Auditory stimuli	1	2	3	4	5
Loud noises/voices	1	2	3	4	5
Strange dogs/cats	1	2	3	4	5
Containment/crate	1	2	3	4	5
Leash/slip collar	1	2	3	4	5
Sudden movements/hand	1	2	3	4	5
Manipulation of feet/tail	1	2	3	4	5
Umbrella	1	2	3	4	5
Falling baggage	1	2	3	4	5
Clip board drop	1	2	3	4	5
Other	1	2	3	4	5

Environmental Reaction Mean Rating: (Must achieve a mean rating of 4 or above)

Part D: Footing

Footing Stimuli or Situations	Poor	Fair	Average	Good	Excellent
Moving conveyor belt	1	2	3	4	5
Stairs	1	2	3	4	5
Tile	1	2	3	4	5
Wire mesh	1	2	3	4	5
Other	1	2	3	4	5

Footing Reaction Mean Rating: (Must achieve a mean rating of 3 or above)

Part E: Obstacles

Obstacles Stimuli or Situations	Poor	Fair	Average	Good	Excellent
Baggage	1	2	3	4	5
Trollies/carts	1	2	3	4	5
Natural objects	1	2	3	4	5
Other	1	2	3	4	5

Obstacles Reaction Mean Rating: (Must achieve a mean rating of 3 or above)

Section 2: General Impression

Rate your general impression of the dog for each item in Section 2. Use a scale of 1 to 5 where the rating of 1 means not at all and 5 means very great degree. **Circle** the number that represents your rating.

General Impression Items	Not At All	Very Little	Some Degree	Great Degree	Very Great Degree
Does the dog make eye contact?	1	2	3	4	5
Will/does the dog make body contact?	1	2	3	4	5
Is it apparent that the dog has had previous training?	1	2	3	4	5
Is it apparent that the dog has had negative conditioning?	1	2	3	4	5
Is the dog sensitive to pain?	1	2	3	4	5
Is the dog curious?	1	2	3	4	5
Is the dog nervous?	1	2	3	4	5
Will the dog fetch?	1	2	3	4	5
Does the dog startle?	1	2	3	4	5
Other	1	2	3	4	5

Comments
Explain dog's recovery time. If the dog is startled, it should take no longer than 3-5 seconds to recover. Note if any technique was used to aid in the recovery process.

_____

_____

_____

_____

_____

_____

_____

Contact the RCPC and the NDDTC for instructions to proceed.

Health Evaluation Protocol Worksheet

Important

The health evaluation must be conducted in the sequence presented on this worksheet by an accredited and licensed veterinarian. Usually, the RCPC accompanies the dog to the veterinarian's office for the health evaluation.

The dog may be eliminated at any point during the health screening process if the results indicate abnormalities.

Dog's Name: _____ Evaluation Date: _____

Sex: M   F   Age: _____ Evaluation Location: _____

Section 1: General Exam
After the general exam is completed, have the veterinarian initial the statement that the findings are within normal limits. Contact the NDDTC before proceeding to Section 2.

Ears/Skin: _____ Weight: _____

Eyes: _____ Heart/Lungs: _____

Coat Condition: _____ Teeth: _____

Any coughing? _____

Any noticeable abnormalities? _____

Is the dog spayed or neutered? Yes _____ No _____ Results are within normal limits _____

Section 2: Heartworm Test (Occult)
Request that the veterinarian perform an occult heartworm test. If the results are within normal limits, have the veterinarian initial the statement below. Contact the NDDTC before proceeding to Section 3.

Date Done: _____ Results are within normal limits _____

Section 3: Blood Test
Request that the veterinarian perform pre-surgical blood work or a blood test that includes the values listed. If the results are within normal limits, have the veterinarian initial the statement below. Contact the NDDTC before proceeding to Section 4.

Kidney values: _____ Liver values: _____

Creatinine: _____ Blood urea nitrogen (BUN): _____

Complete blood count (CBC): _____ Alanine transferase (ALT): _____

Total Protein: _____ Glucose: _____

Results are within normal limits

Section 4: X- Rays

Request that the veterinarian perform ventro-dorsal x-ray of the hips and lateral x-ray of the spine. Have the veterinarian agree to and sign the Statement by Veterinarian. If the results are within normal limits, have the veterinarian initial the statement below. Contact the NDDTC before proceeding to Section 5.

X-rays must be taken in accordance with positioning guidelines set out by American Veterinary Medicine Association (AVMA).

The dog will have to be anesthetized to perform the x-rays.

Ensure that x-rays are clearly marked LEFT or RIGHT.

Send the x-rays to the NDDTC for final approval.

Ventro-dorsal pelvic x-ray: Date done:_____   Results are within normal limits _____

Lateral thoracic-lumbar junction spinal x-ray: Date done:_____   Results are within normal limits _____

Statement by Veterinarian

I consent to retake the x-rays at my own expense if they do not meet the standards set forth by the National Detector Dog Training Center.

Examining Veterinarian:
_____

If you have questions regarding this, please contact the National Detector Dog Training Center (407-816-1221) for clarification.

Section 5: Eating Habits

Request that the veterinarian perform an evaluation of the dog's eating habits to determine if there is evidence of kennel stress. Ask the veterinarian to initial the statement below. If the dog is enthusiastic about food, proceed to Section 6.

1. Was the food consumed:

    A. Quickly _____

    B. Slowly _____

2. Was any food left over?

    A. Yes _____

    B. No _____

Dog does not appear to show stress beyond normal, acceptable limits _____
(Initials)

Section 6: Vaccination

If the overall results of the health evaluation are within normal limits, have the veterinarian administer the following vaccines or record the date when they were given

Vaccinations	Date Given
Rabies (one-year vaccine)	_____
Serial Number: _____	
Producer: _____	
K or MLV: _____	
DHLPP (distemper, hepatitis, leptospirosis, parainfluenza, parvo virus)	_____
Corona	_____
Bordetella (intra nasal)	_____
Fecal exam (internal parasites)	_____

NOTICE The NDDTC does not require a urinalysis or the Lyme disease vaccine; therefore, do not request these.

Contact the RCPC and the NDDTC for instructions to proceed.

United States Department of Agriculture

Marketing and Regulatory Programs

Animal and Plant Health Inspection Service

Plant Protection and Quarantine

Subject: Final Release Form

I _____ do hereby relinquish any legal claim and/or
(Owner)

ownership that I have for _____
(Name of Dog)

_____
(Breed of Dog)

by donation/sale to the U.S. Department of Agriculture for use as a Working Detector Dog. I furthermore understand that this dog will be adopted by the public through U.S. Government procedures upon the retirement of the dog from active duty.

Owner's signature:_____

Date: _____

Location: _____

I _____ request that I be given first right of refusal to the above mentioned dog in the event that it does not pass the training program. I understand I will be responsible for all expenses associated with the dog's return to my possession.

Owner/Agent

Date:

Airline Flight Tracking Worksheet

Important

Do not ship dogs on weekends or holidays without pre-approval from the NDDTC.

Do not use Acepromazine on any dog being shipped to the NDDTC.

Bathe the dog before shipping.

❏ Arriving at NDDTC from _____
(City)

❏ Departing the NDDTC for _____
(City)

Section 1: Detector Dog Information

1. Name: _____
2. Temperature: _____
3. Date and time: _____
4. Contact person: _____
   A. Telephone: _____
   B. Address: _____
   _____
   _____

Section 2: Departure Information

1. Date and time: _____
2. Drop-off time: _____
3. Airline: _____
4. Telephone: _____
5. Cargo: _____ Priority Parcel: _____
6. Flight number: _____
7. First connecting city: _____
   A. Connecting time: _____
   B. Connecting flight number: _____

Section 3: Destination Information

1. City of final destination: _____
2. Time: _____

Tracking Record and Feedback Worksheet

Section 1: Instructions
If you wish to monitor the status of the dog's evaluation and training at the NDDTC, fill out Section 2 of this worksheet and attach it to the Temperament Evaluation Worksheet.

Section 2: Canine Officer
To be completed by the Canine Officer who initially evaluated the dog.

Canine Officer/Team: _____

Address: _____

_____

_____

Dog's Name: _____

Date: _____

Adopted from (shelter, private, etc.): _____

Contact Person: _____

Address of shelter, private donator, etc.:

_____

_____

> Do not write below this line. Section 3 is to be completed by the NDDTC staff.

Section 3: NDDTC Staff
To be completed by the NDDTC Staff.

1. Medicals
    A. Reviewed by:_____
    B. Date reviewed:_____
    C. Approved or disapproved:_____

2. Second Temperament Test Results
   A. Passed.
   B. Failed. If the dog failed the second temperament test, list specific reasons below
3. Protocol Training Results
   A. Passed.
   B. Failed. If the dog failed protocol training, list specific reasons below.
4. Dog Placement
   A. Dog's name: _____
   B. Alias: _____
   C. Assigned to: _____
   D. Location: _____
   E. Date: _____

Reasons Candidate Dog Failed NDDTC Evaluation or Training

_____
_____
_____
_____
_____
_____
_____
_____
_____
_____
_____
_____
_____

AGRICULTURAL DETECTOR DOG TRAINING RECORD

Handler:
Date:
Canine:
Weight:
Port:

Exercise Type	Con/Cnt/Pl	Concentration High Medium Low	Container Hard Medium Soft	Placement High Low Concealed	Scoring + Positive Response − Non-Response | False Response	(+) Handler Assist i Handler Cue (−) Handler Error	
		TARGET ODORS				NON-TARGET ODORS	
						Response	Total Trials

REMARKS

Request to Procure Canines

I, _____ , request that I be considered
(Print your name)
to procure canines for the Detector Dog Program.

I comply with the following:

1) I have at least 3 years experience as a Canine Officer.
2) I have maintained an 80% or better proficiency level with my canine.
3) I have maintained a fully successful evaluation as a Canine Officer.
4) I have been given the permission to procure canines when time allows by my Port Director and Supervisor.

I understand that procurement training does not guarantee that I will be placed on the procurement list and if placed on the list, I may be removed at anytime if the procurement guidelines and procedures are not followed properly. I also understand that I must pass annual validation, if given.

_____   _____
(Signature of Canine Officer)   (Date)

I undersign this to believe this to be true to the best of my knowledge.

Port Director_____ Supervisor_____
                (Print name)   (Print name)

_____   _____
        (Signature and date)   (Signature and date)

* The original signed and completed copy goes to your assigned RCPC.

Statistical Summary - Cargo and Border Canine Operations

Port: Month:

 Team:

Date	Warehouse Dry Goods	Warehouse Refrigerated	Containers Refrigerated	Containers Dry Goods	Rail Containers	Pit Bags Ramp	Couriers Packages	Mail Packages	Private Vehicles	Commercial Vehicles	Buses	Pax Screened	Total Response	Positive Response	Response w/seizure	QMIs Plant	QMIs Animal	Animal Wgt (kg)	Violation	Penalty $	Remarks
1																					
2																					
3																					
4																					
5																					
6																					
7																					
8																					
9																					
10																					
11																					
12																					
13																					
14																					
15																					
16																					
17																					
18																					
19																					
20																					
21																					
22																					
23																					
24																					
25																					
26																					
27																					
28																					
29																					
30																					
31																					

Days work	Warehouse Dry Goods	Warehouse Refrigerated	Containers Refrigerated	Containers Dry Goods	Rail Containers	Pit Bags Ramp	Couriers Packages	Mail Packages	Private Vehicles	Commercial Vehicles	Buses	Pax Screened	Total Response	Positive Response	Response w/seizure	QMIs Plant	QMIs Animal	Animal Wgt (kg)	Violation	Penalty $

Proficiency	Seizure Rate

Index

A

Acknowledgement of receipt, agreement and waiver of liability, sample of 2-6-4

Actions, symptoms of illness or injury 3-4-6

Activity goals 1-1-6

Administering medication
 capsules & tablets 3-4-23
 ear drops & ointment 3-4-27
 eye drops 3-4-26
 eye ointment 3-4-25
 liquids 3-4-24

Aggressive dog, criteria for retiring 2-6-3

Agreement with U.S. Customs Service 2-7-7

Air-scenting, definition of 5-1-1

Alanine transferase, definition of 5-1-1

Alert response, definition of 5-1-1

Anatomy of a beagle 3-1-1
 external 3-1-2
 internal 3-1-3

Animal Welfare Act (AWA) 5-1-1

APHIS
 detector dog agreement with U.S. Customs Service 2-7-7
 key contacts A-1-1
 PPQ key contacts A-1-1 to A-1-6

Assault on detector dog 2-2-1

Attitude, symptoms of illness or injury 3-4-6

AWA (Animal Welfare Act) 5-1-1

B

Baggage information data (PPQ Form 277) H-1-5
 purpose H-1-5

Basic Canine Officer Training
 components 4-1-2
 requirements 4-1-2

Basic kenneling requirements 2-3-2
 checklist for 2-3-3

BCOT
 components 4-1-2
 requirements 4-1-2

Biological break, definition of 5-1-1

Blank container, definition of 5-1-1

Bleeding, signs of and first aid for 3-3-3

Bloating
 how to prevent 3-3-4
 signs of and first aid for 3-3-4

Blood urea nitrogen, definition of 5-1-1

Body, symptoms of illness or injury 3-4-6

Boxes and contents for training D-1-15

C

Canine infectious tracheo-bronchitis (CITB) 3-2-5

Canine interception log for significant pests
 distribution of H-1-5
 purpose H-1-3
 sample of a H-1-4

Canine Officer Supervisor's Training 4-1-11

Index

Canine officers (dog handlers)
  convention in the manual 1-1-7
  roles & responsibilities B-1-2 to B-1-4
    managing detector dogs B-1-2
    public awareness & communication B-1-3
    reporting & documentation B-1-3
    training detector dogs B-1-3
  section 4-1-1

Capsules, steps for administering 3-4-23

Care, general, of detector dogs 3-4-1 to 3-4-28

Caution, advisory symbol 1-1-7

CBP Directive No. 3340-025B, H-1-10 to H-1-15

CITB (canine infectious tracheo-bronchitis) 3-2-5

Claim for Damage, Injury, or Death, Standard Form 95, sample of 2-2-6

Cleaning ears, basic steps for 3-5-5

Coat, symptoms of illness or injury 3-4-5

Cold injury, signs of and first aid for 3-3-5

Collars D-1-1 to D-1-4
  Elizabethan D-1-2
  leather D-1-6
  Martingale D-1-3
  slip D-1-4

Commissioner's situation room reporting, H-1-10 to H-1-15

Complete blood count, definition of 5-1-1

Congressional calls, how to respond 2-5-3

Contacts
  APHIS 1-1-3
  for incidents 2-2-9
  for Plant Protection & Quarantine A-1-1 to A-1-6

Contraband, definition of 5-1-1

Conventions
  bullets 1-1-7
  canine officer 1-1-7
  caution 1-1-7
  chapter 1-1-7
    table of contents 1-1-8
  control data 1-1-8
  heading levels 1-1-8
  highlighting tables, figures, sections 1-1-8
  important note 1-1-8
  must 1-1-8
  notice 1-1-8
  numbering scheme 1-1-9
  section 1-1-9
  tab colors 1-1-9
  warning 1-1-9

Corona virus 3-2-5

Correct response, definition of 5-1-1

Correcting errors in the manual I-1-4

COST 4-1-11

Co-workers (nonhandlers), roles & responsibilities B-1-4

Crates D-1-7
  pads D-1-8

Criteria for retiring detector dogs 2-6-1 to 2-6-3

Cuing dog, definition of 5-1-1

D

Daily health check, guide for 3-4-4 to 3-4-6

Daily transporting detector dogs E-1-1

DDTC 2-1-1

Demonstrations for public awareness 2-5-2

Detector dogs
  daily health check 3-4-4 to 3-4-6
  general care of 3-4-1 to 3-4-28
  shipping E-1-2 to E-1-4
    after shipping procedures E-1-4
    before shipping procedures E-1-2
    during shipping procedures E-1-3
  transporting daily E-1-1

Developmental assignments 2-7-6

Diplomats, courtesy in clearing 2-7-7

Diseases
  infectious 3-2-2 to 3-2-7
    canine infectious tracheo-bronchitis (CITB) 3-2-5
    corona virus 3-2-5
    distemper 3-2-2
    hepatitis 3-2-3
    leptospirosis 3-2-3
    lyme disease 3-2-3
    parvo virus 3-2-4
    prevention and vaccination 3-2-2
    rabies 3-2-4
    Rocky Mountain spotted fever 3-2-4
    summary of 3-2-6 to 3-2-7
  noninfectious 3-2-8

Distemper 3-2-2

Documentation & reporting
  baggage information data (PPQ Form 277) H-1-5
  distribution of operational reports H-1-5
  health care records H-1-10
  narrative report H-1-3
  statistical summary H-1-2
  training records H-1-6

Dog bites, action to take 2-2-4 to 2-2-5

E

Ear(s)
  drops & ointment, administering 3-4-27
  how to clean 3-5-5
  symptoms of illness or injury 3-4-4

Ectoparasite, definition of 5-1-1

Elizabethan collar D-1-2

Emergency care and first aid 3-3-1 to 3-3-11
  bleeding, signs of and first aid for 3-3-3
  bloating, signs of and first aid for 3-3-4
  cold injury, signs of and first aid for 3-3-5
  fractures 3-3-6
    how to splint 3-3-6
  objects in mouth 3-3-5
    signs of and first aid for 3-3-5
  overheating 3-3-7
    signs of and first aid for 3-3-7
  physical restraint 3-3-2
  poisoning, first aid for 3-3-8
  related topics in the manual 3-3-1
  seizures, first aid for 3-3-10
  shock, signs of and first aid for 3-3-11

Emergency response plan for incidents of civil disturbances and natural disasters 2-2-8

Endoparasite, definition of 5-1-1

Equipment D-1-1 to D-1-16
  collars D-1-1 to D-1-4
    Elizabethan D-1-2
    leather D-1-6
    Martingale D-1-3
    slip D-1-4
  crates/portable kennels D-1-7
    crate pads D-1-8
  field supplied 2-1-2
  first aid kit D-1-8
  grooming kit D-1-12
  jackets D-1-13
  leashes (regular & retractable) D-1-13
    illustration of D-1-14
  NDDTC supplied 2-1-1
  refrigerators D-1-14
  reward pouch D-1-14
  suitcases, boxes, & contents D-1-15
  vehicles D-1-16

Exercise, definition of 5-1-2

Extended leave policy 2-7-6

External anatomy of a beagle 3-1-2

External parasites 3-2-8 to 3-2-10
  fleas 3-2-8
  lice 3-2-9
  mites 3-2-9
  summary of 3-2-16
  ticks 3-2-9

Eye(s)
  drops, steps for administering 3-4-26
  ointment, steps for administering 3-4-25
  symptoms of illness or injury 3-4-4

F

False response, definition of 5-1-2

Feet, symptoms of illness or injury 3-4-5

Field supplied equipment and supplies, list of 2-1-2

Figures, highlighting convention 1-1-8

Index

First aid and emergency care 3-3-1 to 3-3-11
  bleeding, signs of and first aid for 3-3-3
  bloating, signs of and first aid for 3-3-4
  cold injury, signs of and first aid for 3-3-5
  fractures 3-3-6
    how to splint 3-3-6
  objects in mouth, signs of and first aid for 3-3-5
  overheating
    signs of and first aid for 3-3-7
  physical restraint 3-3-2
  poisoning, first aid for 3-3-8
  related topics in the manual 3-3-1
  seizures, first aid for 3-3-10
  shock, signs of and first aid for 3-3-11

First aid for
  bleeding 3-3-3
  bloating 3-3-4
  cold injuries 3-3-5
  fractures 3-3-6
  objects in mouth 3-3-5
  overheating 3-3-7
  poisoning 3-3-8
  seizures 3-3-10
  shock 3-3-11

First aid kit D-1-8

First aid kit supplied by NDDTC D-1-9

Fleas 3-2-8

Food guarding, definition of 5-1-2

Fractures
  first aid for 3-3-6
  how to splint 3-3-6

G

General care 3-4-1 to 3-4-28
  administering medication 3-4-23 to 3-4-27
    capsules & tablets 3-4-23
    ear drops & ointment 3-4-27
    eye ointment & drops 3-4-25
    liquids 3-4-24
  cleaning ears, basic steps for 3-5-5
  daily health check 3-4-4 to 3-4-6
  selecting veterinary services 3-4-28

Genital area, symptoms of illness or injury 3-4-5

Goals, activity 1-1-6

Grooming kit D-1-12
  tools for D-1-12

Guidelines agreed to between APHIS and Customs 2-7-7

H

Handbaggage, definition of 5-1-2

Handler error, definition of 5-1-2

Hasty muzzle, definition of 5-1-2

Health care records H-1-10
  distributing H-1-10
  purpose H-1-10
  when and how to maintain H-1-10

Health check, daily, guide for 3-4-4 to 3-4-6

Heartworms 3-2-11
  life cycle 3-2-12

Hepatitis 3-2-3

History of USDA-APHIS detector dogs 1-1-5

Hookworms 3-2-12
  life cycle of 3-2-13

How to splint a fracture 3-3-6

Hyperthermia, definition of 5-1-2

Hypothermia
  first aid and emergency care 3-3-7

I

Illnesses, severe 3-2-8

Implementation process, steps to take for a new detector dog team 2-7-2

Important infectious diseases, summary of 3-2-6 to 3-2-7

Incidents
  assaults 2-2-1
  contacts for 2-2-9
  dog bites 2-2-1
  emergency response plan 2-2-8
  injury and sudden illness 2-2-8

Infectious diseases 3-2-2 to 3-2-7
   canine infectious tracheo-bronchitis (CITB) 3-2-5
   corona virus 3-2-5
   distemper 3-2-2
   hepatitis 3-2-3
   leptospirosis 3-2-3
   lyme disease 3-2-3
   parvo virus 3-2-4
   prevention and vaccination 3-2-2
   rabies 3-2-4
   Rocky Mountain spotted fever 3-2-4
   summary of 3-2-6 to 3-2-7

Injury and illness, action to take 2-2-8

Intermediate host, definition of 5-1-2

Internal anatomy of a beagle 3-1-3

Internal parasites 3-2-11 to 3-2-15
   heartworms 3-2-11
      life cycle of 3-2-12
   hookworms 3-2-12

Internal parasites (continued
   hookworms (continued)
      life cycle of 3-2-13
   roundworms 3-2-13
   summary of 3-2-17
   tapeworms 3-2-14
      life cycle of 3-2-15
   whipworms 3-2-15

J

Jackets D-1-13

K

Kenneling requirements
   basics 2-3-2
   selection of 2-3-1

Kennels
   basic requirements for 2-3-2
      checklist for 2-3-3
      sanitation 2-3-4
   monitoring services 2-3-5
   portable D-1-7
   sanitary requirements 2-3-4
   selecting 2-3-1

Key contacts for APHIS 1-1-3, A-1-1 to A-1-8

L

Latch, secure, definition of 5-1-4

Leashes (regular & retractable) D-1-13
   illustration of D-1-14

Leather collar D-1-6

Legs, symptoms of illness or injury 3-4-5

Leptospirosis 3-2-3

Lice 3-2-9

Life cycle of
   heartworms 3-2-12
   hookworms 3-2-13
   tapeworms 3-2-15

Liquids, steps for administering 3-4-24

Lyme disease 3-2-3

M

Mailing list for the manual, adding & changing I-1-3

Maintaining the manual I-1-1 to I-1-4

Managers, roles & responsibilities B-1-4

Manual
   correcting errors in I-1-4
   mailing list for, adding & changing I-1-3
   maintaining the I-1-1 to I-1-4
   ordering I-1-3
   purpose 1-1-1
   related documents 1-1-7
   revisions I-1-1
   scope 1-1-1
   suggesting improvements for I-1-4
   users 1-1-3
   what it does not cover 1-1-3

Martingale collar D-1-3

Media relations workshop 2-5-5

Media, major, calls
   how to respond to 2-5-3
   tips for positive communications 2-5-3
   workshop for 2-5-5

Mites 3-2-9

Mixed odor, definition of 5-1-2

Index

Mixed target odor, definition of 5-1-2

Monitoring kenneling services 2-3-5

Mouth, symptoms of illness or injury 3-4-4

N

Narrative report H-1-3
　distribution of H-1-5
　purpose H-1-3

National Animal Poison Control Center (NAPCC)
　telephone numbers 3-3-8
　Web site address 3-3-8

National Detector Dog Instructors, roles & responsibilities B-1-9

National Detector Dog Program Manager (NCPM) 5-1-2
　roles & responsibilities B-1-11

National Detector Dog Training Center (NDDTC) 5-1-2
　address for A-1-1

NDDPM. See National Detector Dog Program Manager (NCPM).

NDDTC. See National Detector Dog Training Center (NDDTC).

Nonhandlers (Co-workers), roles & responsibilities B-1-4

Noninfectious diseases and severe illnesses 3-2-8

Nontarget odor(s), definition of 5-1-2

Nose, symptoms of illness or injury 3-4-4

Note, important, helpful hint symbol 1-1-8

Notice, a dangerous situation symbol 1-1-8

O

Objects in mouth, signs of and first aid for 3-3-5

Odor generalization, definition of 5-1-3

Operational reports, distribution of H-1-5

Ordering manuals I-1-3

Outreach information produced by LPA 2-5-2

Overheating
　first aid and emergency care 3-3-7

P

Pads, for crates D-1-8

Parasites 3-2-8 to 3-2-17
　definition of 5-1-3
　external 3-2-8 to 3-2-10
　　fleas 3-2-8
　　lice 3-2-9
　　mites 3-2-9
　　summary of 3-2-16

Parasites (continued)
　external (continued)
　　ticks 3-2-9
　internal 3-2-11 to 3-2-15
　　heartworms 3-2-11
　　　life cycle 3-2-12
　　hookworms 3-2-12
　　　life cycle of 3-2-13
　　roundworms 3-2-13
　　summary of 3-2-17
　　tapeworms 3-2-14
　　　life cycle of 3-2-15
　　whipworms 3-2-15

Parvo virus 3-2-4

Penis, symptoms of illness or injury 3-4-5

Personnel B-1-1

Physical restraint, for first aid and emergency care 3-3-2

Pinpoint, definition of 5-1-3

Placing retired dogs
　options for 2-6-3
　receipt, agreement and waiver of liability 2-6-3

Plant Protection & Quarantine
　key contacts A-1-1 to A-1-6
　　eastern region A-1-2
　　western region A-1-2
　　work locations A-1-3

Poisoning, first aid for 3-3-8

Port directors, roles & responsibilities B-1-4

Portable kennels D-1-7

Positive communications, tips for 2-5-4

Positive response, definition of 5-1-1

Pouch, reward D-1-14

PPQ Officer Training 4-1-12

Prevention of diseases 3-2-2

Primary residence, definition of 5-1-3

Primary reward, definition of 5-1-3

Procuring detector dogs (to be developed) 2-4-1

Public awareness
 outreach information 2-5-2
 performing demonstrations 2-5-2

Purpose of the manual 1-1-1

R

Rabies 3-2-4

RCPM (Regional canine program manager) 5-1-3
 roles & responsibilities B-1-7

Recovery time, definition of 5-1-3

Rectum area, symptoms of illness or injury 3-4-6

Refrigerators D-1-14

Regional canine program manager (RCPM) 5-1-3
 roles & responsibilities B-1-7

Regional program manager (RPM) 5-1-4

Related documents to the manual 1-1-7

Reporting & documentation H-1-1
 baggage information data (PPQ Form 277) H-1-5
 distribution of operational reports H-1-5
 health care records H-1-10
 narrative report H-1-3
 statistical summary H-1-2
 training records H-1-6

Residual odor, definition of 5-1-4

Restraint. See Physical restraint.

Restrictions for the manual 1-1-2

Retiring detector dogs 2-6-1 to 2-6-4
 criteria for 2-6-1 to 2-6-3
  aggressive dog 2-6-3
  dog's ability to work 2-6-1
  health status and history 2-6-2
  when canine officer leaves position 2-6-2
 placing
  options for 2-6-3
  receipt, agreement and waiver of liability 2-6-3

Revisions to the manual I-1-1

Reward pouch D-1-14

Rocky Mountain spotted fever 3-2-4

Roles & responsibilities 1-1-4, B-1-1
 canine officers (dog handlers) B-1-2 to B-1-4
  managing detector dogs B-1-2
  public awareness & communication B-1-3
  reporting & documentation B-1-3

Roles & responsibilities (continued)
 canine officers (continued)
  training detector dogs B-1-3
 co-workers (nonhandlers) B-1-4
 local managers (supervisors, team leaders, port directors) B-1-4
 National Detector Dog instructors B-1-9
 National Detector Dog Program Manager (NCPM) B-1-11
 regional canine program managers B-1-7

Roundworms 3-2-13

RPM (Regional program manager) 5-1-4

S

Sanitary requirements for a kennel 2-3-4

Saturation point, definition of 5-1-4

Scope of the manual 1-1-1

Screening, approach to passenger baggage 2-7-5

Secondary residence, definition of 5-1-4

Secondary reward, definition of 5-1-4

Sections, highlighting convention 1-1-8

Secure latch, definition of 5-1-4

Seizures, first aid for 3-3-10

Selecting
 a kennel 2-3-1
 veterinary services 3-4-28

Index

Severe illnesses 3-2-8

Shipping detector dogs E-1-2 to E-1-4
    after shipping procedures E-1-4
    before shipping procedures E-1-2
    during shipping procedures E-1-3

Shock, first aid and emergency care for 3-3-11

Significant incident report, H-1-16 to H-1-18

Signs of and first aid for
    bleeding 3-3-3
    bloating 3-3-4
    cold injuries 3-3-5
    objects in mouth 3-3-5
    overheating 3-3-7
    shock 3-3-11

Skin, symptoms of illness or injury 3-4-5

Slip collar D-1-4

Speed trials, definition of 5-1-4

Statement, vision 1-1-6

Statistical summary H-1-2
    distribution of H-1-5
    purpose H-1-2
    standard statistics recorded H-1-2

Suggesting improvements in the manual I-1-4

Suitcases and contents for training D-1-15

Summary of
    external parasites 3-2-16
    internal parasites 3-2-17

Supervisors, roles & responsibilities B-1-4

T

Tables, highlighting convention 1-1-8

Tablets, steps for administering 3-4-23

Tapeworms 3-2-14
    life cycle of 3-2-15

Target odor(s), definition of 5-1-4

TDY assignments 2-7-6

Team leaders, roles & responsibilities B-1-4

Temperature, symptoms of illness or injury 3-4-6

Territorial behavior, definition of 5-1-4

Testing, validation, definition of 5-1-4

Ticks 3-2-9

Tips for positive communications 2-5-4

Tours of duty, scheduling 2-7-3

Training records H-1-6
    purpose H-1-6

Transporting detector dogs daily E-1-1

U

U.S. Customs Service, agreement with 2-7-7

Update record for the manual 1-1-2

Users of the manual 1-1-3

Utilizing detector dogs
    courtesy of the port for diplomats 2-7-7
    developmental assignments 2-7-6
    establishing operating procedures for 2-7-3 to 2-7-4
        approach to screening 2-7-5
        flight selection 2-7-3
        tours of duty 2-7-3
        utilizing down time 2-7-4
    extended leave policy 2-7-6
    guidelines agreed to between APHIS and Customs 2-7-7
    process for implementing 2-7-2
    TDY assignments 2-7-6

V

Vaccination for diseases 3-2-2

Validation testing, definition of 5-1-4

Vehicles D-1-16

Veterinary services, selecting 3-4-28

Vision statement 1-1-6

Vulva, symptoms of illness or injury 3-4-5

W

Warning, advisory symbol 1-1-9

Weight
  rating F-1-1
  table to determine range of the dog's F-1-2

What the manual does not cover 1-1-3

Whipworms 3-2-15

Workshop, media relations 2-5-5

Index

www.ingramcontent.com/pod-product-compliance
Lightning Source LLC
Chambersburg PA
CBHW081211230426
43666CB00015B/2713